# Industrial Water Quality Requirements for Reclaimed Water

*The mission of the Awwa Research Foundation (AwwaRF) is to advance the science of water to improve the quality of life. Funded primarily through annual subscription payments from over 1,000 utilities, consulting firms, and manufacturers in North America and abroad, AwwaRF sponsors research on all aspects of drinking water, including supply and resources, treatment, monitoring and analysis, distribution, management, and health effects.*

*From its headquarters in Denver, Colorado, the AwwaRF staff directs and supports the efforts of over 700 volunteers, who are the heart of the research program. These volunteers, serving on various boards and committees, use their expertise to select and monitor research studies to benefit the entire drinking water community.*

*Research findings are disseminated through a number of technology transfer activities, including research reports, conferences, videotape summaries, and periodicals.*

# Industrial Water Quality Requirements for Reclaimed Water

Prepared by:
**David W. Rommelmann, Steven J. Duranceau, Mary W. Stahl, Christina Kamnikar,** and **Roberto M. Gonzales**
Boyle Engineering Corporation
215 Union Boulevard, Suite 500, Lakewood, CO 80228

Sponsored by:
**Awwa Research Foundation**
6666 West Quincy Avenue, Denver, CO 80235-3098

Published by:

Awwa Research Foundation
Advancing the Science of Water®

IWA Publishing

## DISCLAIMER

This study was funded by the Awwa Research Foundation (AwwaRF). AwwaRF assumes no responsibility for the content of the research study reported in this publication or for the opinions or statements of fact expressed in the report. The mention of trade names for commercial products does not represent or imply the approval or endorsement of AwwaRF. This report is presented solely for informational purposes.

**Published by IWA Publishing, Alliance House, 12 Caxton Street, London SW1H 0QS, UK**

Telephone: +44 (0) 20 7654 5500; Fax: +44 (0) 20 7654 5555; Email: publications@iwap.co.uk;
Web: **www.iwapublishing.com**

AwwaRF report number 90979F. Originally published by AwwaRF for its subscribers in 2003.
IWA Publishing version published 2004.



ISBN 1 84339 8869

# CONTENTS

# TABLES

# FIGURES

# FOREWORD

The Awwa Research Foundation is a nonprofit corporation that is dedicated to the implementation of a research effort to help utilities respond to regulatory requirements and traditional high-priority concerns of the industry. The research agenda is developed through a process of consultation with subscribers and drinking water professionals. Under the umbrella of a Strategic Research Plan, the Research Advisory Council prioritizes the suggested project based upon current and future needs, applicability, and past work; the recommendations are forwarded to the Board of Trustees for final selection. The foundation also sponsors research projects through an unsolicited proposal process; the Collaborative Research, Research Applications, and Tailored Collaboration programs; and various joint research efforts with organizations such as the U.S. Environmental Protection Agency, the U.S. Bureau of Reclamation, and the Association of California Water Agencies.

This publication is a result of one of these sponsored studies, and it is hoped that its findings will be applied in communities throughout the world. The following report serves not only as a means of communicating the results of the water industry's centralized research program but also as a tool to enlist the further support of the nonmember utilities and individuals.

Projects are managed closely from their inception to the final report by the foundation's staff and large cadre of volunteers who willingly contribute their time and expertise. The foundation serves a planning and management function and awards contracts to other institutions such as water utilities, universities, and engineering firms. The funding for this research effort comes primarily from the Subscription Program, through which water utilities subscribe to the research program and make an annual payment proportionate to the volume of water they deliver and consultants and manufacturers subscribe based on their annual billings. The program offers a cost-effective and fair method for funding research in the public interest.

A broad spectrum of water supply issues is addressed by the foundation's research agenda: resources, treatment and operations, distribution and storage, water quality and analysis, toxicology, economics, and management. The ultimate purpose of the coordinated effort is to assist water suppliers to provide the highest possible quality of water economically and reliably. The true benefits are realized when the results are implemented at the utility level. The foundation's trustees are pleased to offer this publication as a contribution toward that end.

Water reuse is an important source of water that varies in quantity and quality. Availability of water can be a critical factor and many customers may find reclaimed water to be less expensive than other sources of water. As demand for water grows, a more integrated approach to matching different water qualities with different uses is imperative. Currently, water requirements for different applications exist in a variety of reports. There is not a single comprehensive document that summarizes known beneficial uses of large volumes of water. This document identifies these industrial and ecological applications and known beneficial uses of large volumes of reclaimed water. In addition, applicable water quality requirements are provided as a means for users to evaluate the feasibility of a reclaimed water program.

Edmund G. Archuleta, P.E.
Chair, Board of Trustees
Awwa Research Foundation

James F. Manwaring, P.E.
Executive Director
Awwa Research Foundation

# ACKNOWLEDGMENTS

The authors of this report are indebted to the following utilities and individuals for their cooperation and participation in this project:

- City of El Paso, TX
- Camrosa Water District, CA
- Las Virgenes Municipal Water District, CA
- City of Apopka, FL
- Denver Water Department, CO
- City of Orlando, FL
- City of Albuquerque, NM
- City of San Buenaventura, CA
- City of Aurora, CO

In addition, the advice of the Project Advisory Committee (PAC)—including Paul E. Shoenberger, P.E., Chief of Engineering and Operations, Central Basin Municipal Water District, Carson California; David York, Reuse Coordinator, Florida Department of Environmental Protection, Tallahassee, Florida; Carl Spangenberg, P.E., Senior Engineer, Irvine Ranch Water District, Irvine, California; and Kenneth Thompson $CH_2M$Hill Consulting Engineers, Englewood, Colorado—and the help of initial AwwaRF project officer, Stephanie Passarelli, are appreciated.

The authors wish to acknowledge the assistance of Linda Williams, who set up the database, and Sheila Plesha and Angela Bolin, who helped with word processing and production of graphics.

# EXECUTIVE SUMMARY

Water is a resource that varies in quantity and quality throughout the world. In the US, demand for potable water is growing, growth that becomes more difficult to meet due to the growth occurring in arid and semi-arid areas and growth occurring in areas with limited fresh water supplies. Additionally, as customers of potable water insist on increasing quality, it becomes more expensive for the water provider to treat and deliver potable water. As such, utilities are turning to innovative approaches for meeting the water demands of their customers. Some utilities are also looking for innovative means of meeting increasingly stringent discharge limitations and criteria. Additionally, as the world and national marketplace become more competitive, companies are increasingly looking for ways to reduce costs and improve the reliability of their processes. A common denominator to meeting all of these goals is through the use of reclaimed water.

One aspect of planning a reclaimed water program is to match the water quality needs of potential customers with the expected water quality of the reclaimed water. Typically, many customers, including very large customers, have water uses that do not need to meet potable water quality standards. This would include uses such as cooling and irrigation. Reclaimed water systems are more appealing, both economically and operationally, to a utility if customers with year-round water demands are part of the system. There has not been a great deal of information available to utilities on the water quality requirements for various industrial, commercial, and environmental water uses.

## RESEARCH OBJECTIVES

Water quality requirements for different reclaimed water applications, where available, exist in a variety of reports. The Awwa Research Foundation (AwwaRF) funded this study to provide a document that summarized water quality requirements for industrial uses of reclaimed water. This study specifically excludes irrigation and groundwater recharge applications of reclaimed water because these uses have been extensively addressed in a wide variety of literature (including journal articles, conferences, and books). Additionally, it is unlikely that the pharmaceutical and microchip industries would use reclaimed water without significant additional treatment because their water quality requirements are too stringent. Therefore, the pharmaceutical and microchip industries were also excluded.

Industrial water quality requirements differ significantly over the broad range of industrial process operations. In many manufacturing plants, production processes and product mixes vary and include the full spectrum of water use. Reclaimed water can be effectively applied to cooling, process water, irrigation and washing.

The objective of this project is to identify water users capable of using large volumes of reclaimed water year-round and to provide a reference manual detailing water quality requirements for these year-round large users. The objectives also included getting the viewpoint of suppliers (utilities), users (customers) and regulators (state agencies).

## APPROACH

The project included an information gathering phase and summarization of the information gathered. Information was gathered through a variety of sources, including literature, surveys of various stakeholders and databases. The surveys included both "check-the-box" features as well as more open-ended questions. Surveys were sent to utilities, state regulatory personnel, and large industrial customers. Surveys were initially completed by early 2001, with follow-up phone calls to elicit additional information and clarification occurring through the spring and summer of 2001. Databases maintained by state agencies were accessed for information. No industry databases, with data addressing either the quantities on non-potable water consumed by a particular industry or application, or the quality standards applied to it, were available through the Internet in Spring of 2001.

## CONCLUSIONS

The standards for water use vary with the geographic area and the types of water quality control problems present. General concerns crossed industry uses and include:

1. Bacterial growth, slime/scale formation, and foaming due to residual organics.
2. Potential corrosion in copper-based alloys.
3. Interference with free chlorine disinfection due to ammonia.
4. Stimulation of microbial growth due to ammonia, nitrates, and phosphorous.
5. Deposition/clogging from suspended solids.
6. Scale-formation from calcium, magnesium, iron, silica, and phosphorous.
7. Staining from iron and manganese.
8. Corrosion from sulfates.

In general, industrial users need a minimum quantity of water with water quality that varies little. Consistency of the water quality, regardless of the level, is as important, or more important than the actual quality of the water. Changes in water quality that upset manufacturing processes or the treatment processes of individual users can cause a great deal of economic damage to a customer. It should be noted that customer treatment processes are many times "set" with little attention and few feedback loops for adjustment of chemical feed.

Typically, water quality for key parameters will not vary substantially within an industry. Note, however, that many facilities have water treatment equipment that can accept varying influent concentration. Because of this, the water supplier can negotiate actual water quality supplied to an industrial facility. In addition, the supplier should require that the industrial employer is responsible for insuring appropriate and safe use of reclaimed water in the facility.

The experience of the authors and the lack of feedback from the surveys suggest that customers, even large customers whose products are heavily dependent on water, do not have a thorough understanding of water chemistry. It may be very likely that a potential customer of a reclaimed system may not know the minimum water quality for a particular use, beyond one or two key parameters. Rather, the user understands what has worked to date, based on historical water supplies, but may not know the effects of any particular changes. The surveys used the benchmark of drinking water quality for the following reasons:

**Table ES.1**
**Water quality ranges for industrial processes**

| Parameter* | Pulp & paper | Chemical | Petro-chemical | Textiles | Recirculating cooling systems |
|---|---|---|---|---|---|
| Fe | 0.1-1.0 | 0.1 | 1.0 | 0.1-0.3 | 0.5 |
| Mn | 0.05-0.5 | 0.1 | — | 0.01-0.05 | 0.5 |
| Cl | 200-1000 | 500 | 300 | — | 500 |
| Hardness | 100 | 250 | 350 | 25 | 650 |
| Alkalinity | — | 125 | — | — | 350 |
| TDS | — | 1000 | 1000 | 100 | 500 |
| TSS | 10 | 5 | 10 | 5 | 100 |
| Color | 10-30 | 20 | — | 5 | — |
| pH | 6-10 | 6.2-8.3 | 6-9 | — | 6.9-9.0 |

*Source:* Water Pollution Control Federation, 1989 and USEPA Guidelines for Water Reuse, 1992.
* All values in mg/L: except color and pH. "—" indicates no information was provided in the reference.

1. Many industrial managers are not aware of specific water quality needs.
2. Drinking water quality is somewhat consistent across the country.

Note that other benchmarks were considered, but were not chosen because they lacked consistency across the country. For example, raw water can have turbidities ranging from less than 2 to greater than 100. Another constituent of concern for many industrial uses, iron, raw water values can range from non-detect to 1 mg/L or greater. It should also be noted that many industries may not know their limits for particular constituents. It has been the authors' experience that industrial facilities "make-it-work."

Ranges of industrial water quality requirements for various parameters are shown in Table ES.1.

The pulp and paper industry uses water to make pulp, washing, transportation, refinery of products, and cooling. There are different grades of paper, with increasing brightness requiring increasingly better water quality. Specific constituents of concern include suspended solids, phosphates, iron, manganese, and microorganisms.

Water quality requirements for the chemical industry vary widely with the product being manufactured. In general, desired water quality is neutral pH, moderately soft, low turbidity, low suspended solids, and low silica.

The textile industry has similar concerns as the pulp and paper industry, with staining being a primary concern. General water quality requirements are low turbidity, color, iron, and manganese. Nitrates and nitrites may interfere with the dying processes.

The petroleum industry generally tolerates water of relatively poor quality.

Reclaimed water has been used for cooling purposes for many decades. Cooling water is a common need across all industries and the water quality required is dependent on the type of cooling process rather than the industry. Cooling systems include range from dipping tanks to re-circulating heat exchangers. Cooling systems that are once through have lower water quality requirements than re-circulating systems. Water quality issues for cooling systems include scaling, corrosion, biological growths, and fouling. Raw water is frequently used in cooling systems; reclaimed water can frequently be used with little or no additional treatment.

Boiler feed water is generally of higher quality than potable water. The level of purity is dependent on the boiler system. As most users already perform additional treatment, reclaimed water can many times be used as feed water to the individual treatment system.

Stream augmentation and other aesthetic water features have varying water quality requirements, typically dictated by the National Pollution Discharge Elimination System permitting process.

The survey responses of reclaimed water utilities indicate that few have customers requesting specific water quality standards. Reclaimed water sources for utilities include stormwater runoff, graywater sources (in-house wastewater, from sources other than toilets), and treated wastewater. Many utilities are planning upgrades and expansions to reclaimed water facilities to meet a growing demand for reclaimed water.

The states have varying requirements for reclaimed water systems. The state regulations that were examined for this study had the purpose of primarily protecting public health and secondarily protecting the environment. Most states have classifications of use (typically restricted and unrestricted) although there is not consistency between states on the classification definitions. Many states have a case-by-case basis for water quality, monitoring, and volumes applied for many kinds of industrial and environmental reclaimed water application when they are granting a reclaimed water permit.

## RECOMMENDATIONS

This study is the starting point for a utility planning a new or expanded reclaimed water system. However, water quality requirements for industrial users appear to be site specific. Thus, utilities will need to contact potential users individually and discuss their specific water quality needs. Additionally, in many states, industrial uses are not regulated specifically, and individual state regulators will need to be contacted for each project. This need to initiate one-on-one conversation should not discourage the planners of reclaimed water systems as industrial users often represent the desired year-round base-loading that makes a project economical. The communication with customers can also be used as an opportunity to enhance an existing or new relationship. The planners of reclaimed water systems should recognize that a key limitation of industrial customers is the need to provide consistent water quality and quantity.

## FUTURE RESEARCH

As industrial users of reclaimed water systems become more common, this area of study should be revisited. Additionally, research on consistent regulatory standards is needed as well as regulatory standards that account for reclaimed systems that are above and beyond irrigation.

# CHAPTER 1
# INTRODUCTION

## PROJECT BACKGROUND

Water is a resource that varies in quantity and quality throughout the world. As demand for water grows, quality typically deteriorates, and its availability may decrease. As the availability of water has decreased, utilities have turned to innovative approaches for serving their customer's need, including implementing the concept of reuse. Utilities have also turned to the concept of reuse as an innovative method of meeting discharge limitations and criteria set by regulatory agencies. Successful reclaimed water programs have used an integrated approach to matching different water qualities with different water uses. The following definitions will be used in this report:

Reuse – The use of reclaimed water for non-potable purposes.
Reclaimed Water – Water that has been treated to state standards for discharge from a treatment facility.

Water reuse is more appealing, economically and operationally, to a utility if customers with year-round water demands can be brought onto the system. This baseline demand is often the determining factor in a reclaimed water program's economic viability. Most utilities evaluating the feasibility of a reclaimed water program do not have quick access to water quality requirements for various industrial, commercial, and environmental enhancement (wetlands, water feature, etc.) reclaimed water applications.

Reclaimed water is an important source of water in many areas. Reclaimed water also conserves raw water resources. In many cases, customers may find it to be less expensive than developing new sources of water after costs of potable water development are considered. Typically, reclaimed water has been synonymous with landscape irrigation, which usually provides a small baseline demand that is subject to seasonal fluctuations, thereby decreasing the economic feasibility of many reclaimed water programs. By reducing the peaking effect of landscape irrigation demands through the use of reclaimed water, potable water treatment facility and distribution system capacities can be smaller, or capacity expansions can be reduced. In the past, utilities in areas of the country where there is no chronic shortage of water, rarely consider reclaimed water programs, especially if the predominant use would be for irrigation. Now more and more utilities are considering reclaimed water as a means of better managing their water supply. In many cases, municipal wastewater discharge criteria are such that with little or no additional treatment, the treated wastewater is acceptable for a major customer's use. Several major reclaimed water programs currently in operation in the U.S. have successfully retrofitted major customers with year-round reclaimed water.

Most state permitting agencies deal with non-landscape irrigation reclaimed water systems on a case-by-case basis. The level and type of human and ecosystem exposure must be considered for each specific use. In the case of applications that have no precedent-setting use history, permitting agencies tend to err on the safe side or not permit the use at all. Each state has approached the permitting of reclaimed water in differing ways.

Table 1.1 presents examples of typical reclaimed water applications.

**Table 1.1**
**Reclaimed water applications**

| Application | Examples |
|---|---|
| Agricultural Irrigation | Fodder, Fiber and Seed Crops, Food Crops, Nurseries, Frost Protection |
| Environmental | Streamflow Augmentation, Marshes, Wetlands, Fisheries |
| Groundwater Recharge | Recharge Potable Aquifer, Salt Water Intrusion Control, Subsidence Control |
| Impoundments | Ornamental, Recreational |
| Industrial Processes | Cooling, Boiler Feed, Washing, Stack Scrubbing, Process Water |
| Landscape Irrigation | Parks, Cemeteries, Golf Courses, Thoroughfare Right-of Way, School Grounds, Greenbelts, Residential Lawns |
| Miscellaneous | Aquaculture, Snow-Making, Construction, Dust Control, Livestock Watering |
| Non-Potable Urban | Landscape Irrigation, Fire Protection, Toilet Flushing, Air Conditioning, Vehicle Washing, Street Washing, Ice making/Ice rinks, Commercial laundries |
| Indirect Potable Reuse | Augmenting Surface Waters Used for Public Water Supply |

The focus of this report is the industrial use of reclaimed water. As noted above, reclaimed water for industrial applications can help make a reclaimed water project more economical than those that restrict themselves to irrigation use. Additionally, large users can be served by a single transmission main rather than a distribution rework, assisting in making project economics favorable. Large industrial applications include the following:

- Chemical Production
- Power Plant Cooling
- Pulp and Paper Production
- Cooling
- Textiles Production

## PROJECT NEED

Water quality requirements for different reclaimed water applications, where available, exist in a variety of reports. The goal of this research was to provide a document that summarized water quality requirements for large volumes of reclaimed water. At a minimum, these uses include commercial, industrial, agriculture, and applications for environmental enhancements.

Irrigation and groundwater recharge are not included because they have been previously addressed extensively in other literature.

Industrial water quality requirements differ significantly over the broad range of industrial process operations. In many manufacturing plants, production processes and product mixes vary and include the full spectrum of water use.

Quality requirements depend primarily on type of use and site-specific circumstances. Reclaimed water can be effectively applied to the following industrial water use categories:

- Cooling Water
- Boiler Feed Water
- Process Water
- Irrigation
- Washing

Typically, water quality for key parameters will not vary substantially within an industry. Note, however, that many facilities have water treatment equipment that can accept varying influent concentrations. Because of this, the water supplier can negotiate actual water quality supplied to an industrial facility. In addition, the water supplier should require that the industrial employer is responsible for ensuring appropriate and safe use of reclaimed water in the facility.

## RESEARCH OBJECTIVES

The objective of this project is to identify water users capable of using large volumes of reclaimed water year-round, to identify areas where reclaimed water can be used for environmental enhancement, and to provide a reference manual detailing water quality requirements for these year-round large user and ecological applications.

State regulatory on-line databases for water quality and quantity criteria were accessed for non-landscape irrigation users in Arizona, California, Colorado, Florida, Hawaii, Maryland, Nevada, Oregon, Texas and Utah. Irrigation use lies outside the purview of this reference work, and is not reviewed with respect to water quality requirements. The Water Quality Association and members of the Project Advisory Committee (PAC), as well as utility team members, provided data and specific project listings for possible additional sources of water quality and usage data.

# CHAPTER 2
# METHODS AND MATERIALS

## LITERATURE SEARCH – SOURCES

Government and privately published papers listed in the reference bibliography were consulted for the majority of the tables, charts, and standards cited in this paper. Hard copies of the regulations quoted here were also provided by the State agencies consulted. Where tables were taken from these sources, they have been referenced as such in the text.

American Water Works Association (AWWA) and the University of Central Florida (UCF) Library Information Network & Exchange (LINE) were contacted and requested to search journals, books, government reports, conference proceedings and other types of publications using the following key word searches:

- *"Water Quality Requirements for Recycled/Reclaimed/Reuse Industrial Water"*
- "Recycled/Reclaimed/Reused Water and Requirements/Regulations/Laws and Water Quality and Use/Reuse"

Table 2.1 presents the databases that were searched.

AWWA searched the WATERNET database. The AWWA WATERNET database is comprised of references from journals, in addition to books, government reports and other types of publications. Table 2.2 presents the AWWA references that were searched.

UCF LINE searched 26 DIALOG databases. Of these 26 databases, 7 databases had unique records. Sixty-four records were initially identified.

After reviewing the abstracts and excluding the irrigation and agricultural references 20 citations were references.

## USEPA GUIDELINES

The 1992 USEPA Manual, *Guidelines for Water Reuse* (see Appendix D), contains recommendations applicable to domestic wastewater from municipal or other wastewater

**Table 2.1**
**Contact reference database**

| Contact | Databases |
|---|---|
| AWWA | WATERNET |
| UCF LINE | Water Resources Abstracts<br>Ei Compendex (Engineering Index)<br>Enviroline<br>NTIS<br>Pascal<br>Toxline<br>GeoRef |

**Table 2.2**
**AWWA WATERNET references**

| | |
|---|---|
| Awwa Research Foundation Research News | Journal of the Institution of Water Engineers and Scientists |
| Awwa Research Foundation Water Quality Research News | Journal of the Irrigation and Drainage Division, ASCE |
| Aqua (Journal of Water Supply Research and Technology) | Mainstream |
| Aqua (The Journal of the International Water Supply Association) | Opflow |
| Big Sky Clearwater | Ozone: Science & Engineering |
| Desalination: The International Journal on the Science and Technology of Desalting and Purification | Plumbing Standards |
| Journal American Water Works Association | Science |
| Journal Water Pollution Control Federation | Science et Techniques de l'Eau |
| Journal of Environmental Engineering | Water |
| Journal of Freshwater | Water & Pollution Control |
| Journal of New England Water Works Association | Water Research |
| Journal of Water Resources Planning and Management | Water Resources Bulletin |
| Journal of the Environmental Engineering Division ASCE | Water Science and Technology |
| Journal of the Chartered Institution of Water and Environmental Management | Waterworld News |
| Journal of the Institution of Water and Environmental Management | |

treatment facilities having a limited input of industrial waste. They are based on reclaimed water experience in the U.S. and elsewhere; research and pilot plant or demonstration study data; technical material from the literature; various states' reuse regulations, policies, and guidelines, as of 1992; attainability and sound engineering practice.

Most of the criteria are aimed at preserving public health, and control of pathogenic organisms. Since it would be impractical to monitor water for all organisms, surrogate parameters such as fecal and total coliform count are used instead. The parameters do not include suggested parasite or virus limits, instead using total suspended solids and turbidity as a surrogate measurement.

## SURVEY DEVELOPMENT

As there is not a great deal of literature available on industrial and ecological reclaimed water use, the data collection effort included surveying interested parties to attempt to obtain more detailed information. Three separate surveys were developed for each of the categories of stakeholders, including large customers, utilities and state regulatory agencies. These surveys can be found in Appendix B.

Information that was desired from each survey participant included type of use or industry, water quality requirements, water quantity and use. Each survey participant was asked in a slightly different way to tailor the survey towards his or her viewpoint. For example, large customers were asked for details on what is manufactured and how water is used, utilities were asked to list the types of manufacturing industries that use or plan to use reclaimed water, and state regulators were asked to list the types of manufacturing customers that use reclaimed water within their state, as well as being asked to provide copies of their state regulations.

For the large users, the survey was addressed to personnel not fluent in water quality/water treatment parameters. The experience of the research team is that industrial water users are not well versed in water quality requirements for their particular purposes. An individual manager may be aware of one or two constituents that have posed problems in his/her facility but is not generally aware of other parameters that could be a problem if present in higher concentration. Thus, the categories for water quality included "less than drinking water", "drinking water" and "greater than drinking water." Drinking water standards were used as a standard as it is a common standard whose quality is relatively consistent across the country (at least with respect to Safe Drinking Water Standards). Other benchmarks were considered but were considered more subjective and requiring greater knowledge of water quality and regulations.

AwwaRF and the Project Advisory Committee (PAC) reviewed draft copies of the surveys. Their comments and responses were incorporated into the final versions that were mailed to identified survey participants. Copies of the final surveys are in Appendix B.

Survey participants were identified by a number of methods. Utilities included those utilities providing in-kind services to the project. Additionally, PAC members identified utilities as having a large industrial base or having a unique industrial customer. Large customers were initially identified by utilities participating in the project. PAC members identified additional customers. State agencies were defined in the project scope.

Each identified survey participant was mailed via US mail or e-mailed a cover letter and survey. The cover letters explained the purpose of the project and the desired participation by the respondent. Where surveys were sent to customers of participating utilities, the cover letter was sent by the utility. A map of participating U.S. states is shown in Figure 2.1. The list of survey participants is shown in Table 2.3.

## COMPILATION OF SURVEY RESULTS

The answers from the surveys were entered into a database created in Access 2000, after which queries for the specific quantity and quality characteristics were then constructed (see Appendix B for copies of the surveys, and Appendix C for copies of the database results). Approximately 20 queries comparing potable to non-potable use, peak time of day, and additional required industrial water quality parameters were constructed, with the results exported to Microsoft Excel for formatting. Graphs and charts were created in Excel using the Chart Wizard,

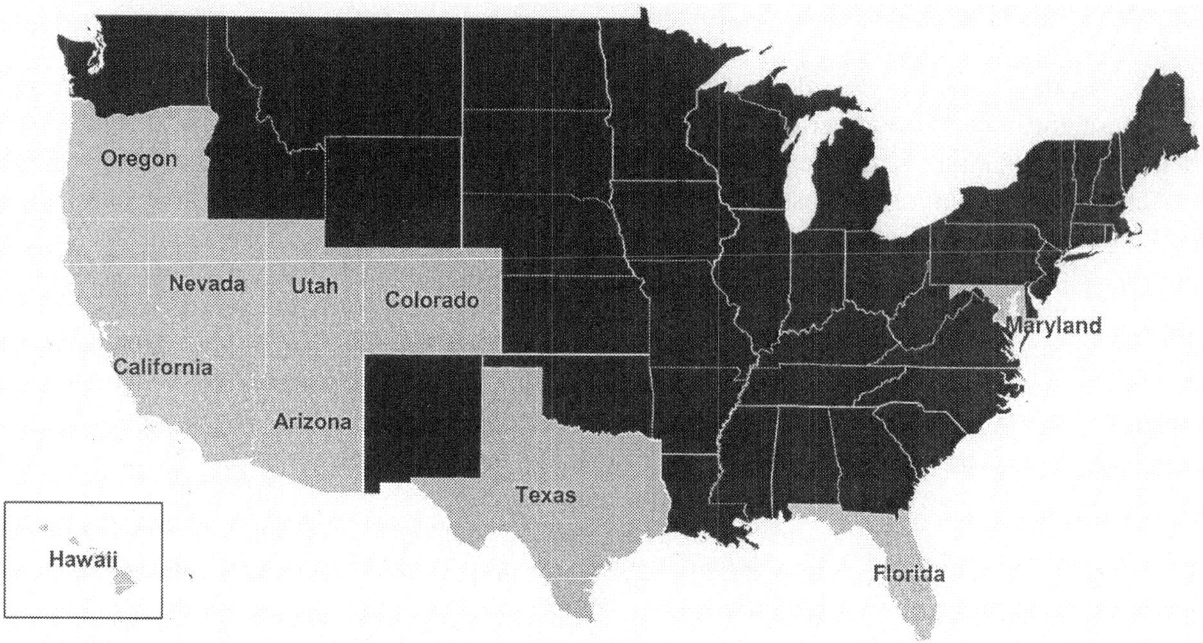


**Figure 2.1 U.S. states with participating utilities**

and were subsequently edited for clarity and completeness. Extra editing of the queries was occasionally required, due to the wide variation in responses to the questions, and some responses were compiled into simpler categories in order to make comparisons across industry categories, or states, easier. In cases where specific data was found to be missing from the database, follow-up phone calls were made to the users or utilities, and further details were sought in order to obtain as much complete information as possible. Phone calls were made to those who had not returned the survey in order to increase the response rate, and the surveys were then completed over the phone with an author. Additionally, calls were made to contacts provided by survey respondents in industry and at municipal utilities, in order to obtain more information regarding water quality requirements.

## DATABASES – STATE & INDUSTRY

At the time of data collection (Spring 2001), California, Florida, Oregon and Texas maintained databases on utilities and recycled water users within their states. Oregon's database was undergoing a restructuring at the time the research was conducted. California's 1998 Water Plan Update, Florida's 1999 Reuse Inventory and Texas's Municipal 1997 Wastewater Reuse database were available at the addresses in Appendix A. Florida's inventory outlines both the potential and actual reclaimed water applied in industrial, agricultural, commercial and irrigation uses by county and city, while Texas's database breaks down the industrial and irrigation volumes of reclaimed water use by county and city. California's 1998 Water Plan Update also contains some data regarding both the current and projected use of recycled water in the state by water district under normal and drought conditions.

**Table 2.3**
**Survey participants**

| Industrial/Ecological Users | |
|---|---|
| Olivas Adobe Vista Parks Department, CA | Westlake Village Inn, CA |
| A.L. Johnson Company, CA | Monroe County Detention Center, FL |
| Allied Fasteners, Inc., CA | Vero Beach Power Plant, FL |
| SCI, CA | Electric Power Research Institute, CA |
| Siemens Solar Industries L.P., CA | Total Petroleum, CO |
| Tierra Rejada Golf Club, CA | Florida Department of Corrections, FL |
| El Paso Electric Company, TX | Sand Creek Chemical, CO |
| The Church of Jesus Christ of LDS, CA | Rocky Mountain Arsenal, CO |
| City of Calabasas HOA, CA | General Mail Facility, CO |
| County Line Shopping Center, CA | Park Hill Golf Course/Clayton College, CO |
| Creekside Calabasas HOA, CA | Gateway Park/Sand Creek Metro District, CO |
| Greenbrier Properties, Inc., CA | Denver International Airport, CO |
| Helmet House, CA | Denver Bulk Mail Center, CO |
| La Ventura Properties, CA | ASARCO, CO |
| Lakeshore Oak View Ranch, Regency Hills, CA | Tallahassee's Purdem Power Plant, FL |
| The Kanan Group, CA | Florida Power, Inc., FL |
| Calabasas Golf and Country Club, CA | Xcel Energy – Cherokee Plant, CO |
| Soka University of America, CA | Florida Department of Corrections, FL |
| South Hills Property Owners Association, CA | Lakeland McIntosh Power Plant, FL |
| Stonecrest Patio Homeowners Association, CA | Royalty Carpets, CA |
| Watergate Property Owners, CA | Dillon Dairy, CO |
| City of Westlake Village, CA | |
| **Utilities** | |
| City of San Diego, CA | City of Tallahassee, FL |
| Las Virgenes Municipal Water District, CA | Denver Water Department, CO |
| City of Apopka, FL | West Basin Municipal Water District, CA |
| Reedy Creek Improvement District, FL | Irvine Ranch Water District, CA |
| City of San Buenaventura, CA | South Bermuda WWTP, FL |
| T.P. Smith, FL | Naples Waste Treatment Plant, FL |
| **State Regulatory Agencies** | |
| Arizona Department of Environmental Quality, AZ | |
| Nevada Department of Environmental Protection, NV | |
| Division of Water Quality Rules, UT | |
| Florida Department of Environmental Protection, FL | |
| Colorado Department of Public Health and Environment, CO | |
| Maryland Department of Environment, MD | |
| Oregon Department of Environmental Quality, OR | |

Arizona, California, Colorado, Florida, Hawaii, Maryland, Nevada, Oregon, Texas and Utah, had websites where their laws and regulations are posted in either HTML or PDF format at the time of data collection (Spring 2001). See the addresses in Appendix A for exact locations of the applicable regulations (current as of Spring 2003). Maryland did not have reclaimed water regulations posted on a website.

No industry databases, with data addressing either the quantities on non-potable water consumed by a particular industry or application, or the quality standards applied to it, were available through the Internet at the time of data collection.

# CHAPTER 3
# RESULTS AND DISCUSSION

## WATER QUALITY BY INDUSTRY

Information on industry standards for water quality is not well developed for a number of reasons. The primary reason is the diversity of manufacturing within each industry as well as the diversity amongst plants. Many previously published guidelines include statements such as the following from the State of New Jersey: "The uniqueness of each industrial reuse application makes it impossible to establish specific treatment standards. Prior to implementation, all industrial reuse systems require a case-by-case review by the Department." This is also reflected in many state regulations that require case-by-case review for industrial uses of reclaimed water. The authors strongly recommend that potential customers for reclaimed water be contacted directly for their specific needs. The standards for water use vary with the geographic area and types of water quality control problems present.

Review, however, of the previously listed literature yielded basic EPA guidelines and other industry-recommended standards in the areas as listed in Table 3.1 and discussed by industry following Table 3.1. Most of these standards are applied not only to protect the public and environmental health, but also to maintain the proper operation of the industrial system in question and to preserve the quality of the products manufactured. These standards can be used as guidelines in the planning stages of a reclaimed water project. General concerns regarding certain parameters cross industries. These concerns are listed in Table 3.1.

**Table 3.1**
**Reclaimed water quality concerns for industrial reuse**

| Parameter | Concern or problem |
|---|---|
| Residual Organics | Bacteria growth, Slime/scale-formation, Foaming in boilers |
| Ammonia | Interferes with formation of free chlorine residual, Causes stress corrosion in copper-based alloys, Stimulates microbial growth |
| Phosphorous | Scale formation, Stimulates microbial growth |
| Suspended Solids | Deposition, "seed" for microbial growth |
| Calcium and Magnesium | Scale-formation |
| Iron | Scale-formation, Staining |
| Nitrate | Stimulates microbial grow |
| Silica | Scale-formation |
| Sulfate | Corrosion |

Adapted from WPCF *Water Reuse Manual of Practice,* 1989.

## Pulp and Paper Industry

According to the California Office of Water Recycling, national quantities of recycled water in the pulp and paper industry were expected to reach 58.3 billion gallons a day, or 11.8% of the total used, by the year 2000. Data substantiating that this projection actually occurred was not available in Spring of 2001. "This increase in recycle will decrease the need for freshwater withdrawals by 36 percent, to 5.4 bgd, and wastewater discharges by 85 percent, to 1.1 bgd [across the nation]." In paper manufacturing processes, water is used for cooling and grinding wood chips to make pulp; repeated washings of the pulp; and the transportation of fibers through the bleaching, refinery and sheet-forming processes. Water is also needed for boiler feed and cooling addressed below, but since the criteria for these processes are fairly consistent across industries, they will be addressed separately.

Water used in pulp and paper manufacturing processes must meet certain chemical and physical criteria, based on the grade of paper being produced. The "brown" grades of paper can tolerate the lowest quality of water in their production, including asphalt or tar-saturated papers, linerboard, low-brightness carton board, packaging and insulating board. The low-brightness "white" grades are the next most forgiving, with newsprint, paper toweling and bathroom tissues, facial tissues, table napkins and "white towels" for kitchen use coming under this heading. Lastly, the "fine papers," those with high whiteness (brightness) grades, are the most sensitive to water quality impurities, and require the most stringent standards in order to prevent uneven tone and the rapid 'aging' of paper due to metals contamination. The efficient fixing of the resins and fibers in paper is also sensitive to water quality, and can be effected by phosphates and other dispersants, as well as metal ions in the water. Iron, manganese, and microorganisms can also discolor water and thus the final product, and a high count of suspended solids can decrease the brightness of paper produced. Biological growths can also affect the texture and uniformity of the paper produced. Tables 3.2 and 3.3 include the summary of specifications for process water quality for the manufacture of various grades of paper across different companies, with comparisons to the municipal water sources available. Table 3.4 includes the requirements for specific manufacturers compared to reclaimed water quantity.

Aside from the physical and chemical requirements for reclaimed water, the consistency of the water quality supplied is also an issue. An upset, failure or accidental release at a wastewater or reclaimed water treatment plant can allow contaminants into the waste stream that can ruin a run of paper at great cost to the manufacturer. Confidence that the water purchased will be of consistent water quality is as important as knowing the water quality level to which reclaimed water will be treated prior to use in manufacturing processes or cooling and boiler systems. Adequate warning systems and notifications of problems at a plant should be in place to reduce the probability of costly accidents for the consumer of reclaimed water.

## Chemical Industry

The water quality requirements for different chemical industries vary widely according to production requirements. Process water is used in manufacturing for product mix, washing and rinsing, transporting product and reaction media. According to the Water Pollution Control Federation (1989), neutral waters (pH ranging from 6.2 to 8.3), moderately soft with low turbidity, low suspended solids and low amounts of silica are required. Dissolved solids and

**Table 3.2**
**Industrial process water quality requirements**

| | Pulp & paper | | | | | Textiles | | |
|---|---|---|---|---|---|---|---|---|
| Parameter* | Mechanical pulping | Pulp & paper, unbleached | Pulp & paper, bleached | Chemical | Petrochemical & coal | Sizing suspension | Scouring, bleach & dye | Semiconductors |
| Cu | — | — | — | — | 0.05 | 0.01 | — | < 0.001 |
| Fe | 0.3 | 1.0 | 0.1 | 0.1 | 1.0 | 0.3 | 0.1 | — |
| Mn | 0.1 | 0.5 | 0.05 | 0.1 | | 0.05 | 0.01 | — |
| Ca | — | 20 | 20 | 68 | 75 | — | — | — |
| Mg | — | 12 | 12 | 19 | 30 | — | — | — |
| Cl | 1,000 | 200 | 200 | 500 | 300 | — | — | < 2 |
| $HCO_3$ | — | — | — | 128 | — | — | — | — |
| $NO_3$ | — | — | — | 5 | — | — | — | — |
| $SO_4$ | — | — | — | 100 | — | — | — | — |
| $SiO_2$ | — | 50 | 50 | 50 | — | — | — | <0.005 |
| Hardness | — | 100 | 100 | 250 | 350 | 25 | 25 | — |
| Alkalinity | — | — | — | 125 | — | — | — | — |
| TDS | — | — | — | 1,000 | 1,000 | 100 | 100 | — |
| TSS | — | 10 | 10 | 5 | 10 | 5 | 5 | — |
| Color | 30 | 30 | 10 | 20 | — | 5 | 5 | — |
| pH | 6-10 | 6-10 | 6-10 | 6.2-8.3 | 6-9 | — | — | — |

*Source:* Water Pollution Control Federation, 1989 and USEPA Guideline for Water Reuse, 1992.
* All values in mg/L: except color and pH; "—" are left where no information was provided in the reference.

chlorine content are not critical for the production of most chemicals. See Table 3.2 for USEPA Guidelines.

### Textile Industry

Textile production requires water that does not stain during the process, and thus the waters used must be low in turbidity, color, iron, and manganese. Process water is used for washing, scouring, dyeing, printing, bleaching and finishing. There are three major categories of textiles: wool, cotton and synthetics. Each requires different water quality for process waters. Hardness can cause problems in some of the wash processes, and nitrates and nitrites may interfere with dying processes, according to respondents to our survey. As mentioned before, a consistent level of water quality is as important as the concentration of various parameters, so that systems adjusted for a certain level of key contaminants are not compromised by an influx of

**Table 3.3**
**Summary of specifications for process water quality for the manufacture of various papers (after McKee & Wolf) (concentration, mg/L)**

| Parameter | Brown grades | | White grades | | |
|---|---|---|---|---|---|
| | Groundwood (tar) paper | Soda and sulfate pulps | Kraft paper (towels) bleached | Unbleached | Fine paper |
| Turbidity in NTU | 70* | 35* | 56 | 140 | 14 |
| Color in platinum units[2] | 30 | 5 | 25 | 100 | 5 |
| Total hardness as $CaCO_3$ | 200 | 100 | 100 | 200 | 100 |
| Calcium hardness as $CaCO_3$ | — | 50 | — | — | 50 |
| Magnesium hardness as $CaCO_3$ | — | 50 | — | — | — |
| Alkalinity to methyl orange as $CaCO_3$ | 150 | 75 | 75 | 150 | 75 |
| Iron | 0.3 | 0.1 | 0.2 | 1.0 | 0.1 |
| Manganese | 0.1 | 0.05 | 0.1 | 0.5 | 0.05 |
| Silica as soluble $SiO_2$ | 50 | 20 | 50 | 100 | 20 |
| Total dissolved solids | 500 | 250 | 300 | 500 | 200 |
| Free carbon dioxide as $CO_2$ | 10 | 10 | 10 | 10 | 10 |
| Chlorides | 75 | 75 | 200 | 200 | — |
| Residual chlorine | — | — | — | — | 2.0 |

*Source:* Office of Water Recycling. Water Recycling in the Pulp and Paper Industry in California, 1982.
* Materials causing turbidity must not be gritty.

water with different or worse water quality values. Water quality parameters are listed in Table 3.2.

### Petroleum and Coal Industry

Processes for the production of petroleum and coal can usually tolerate water of relatively low quality. A pH of 6 to 9 and a moderate suspended solids level of no more than 10 mg/L are typically acceptable. Water quality parameters are listed in Table 3.2.

**Table 3.4**
**Process water requirements for paper production and municipal wastewater treatment plan effluent quality (concentration mg/L unless noted)**

| Parameter (mg/L) | Celotex, L.A. (asphalt roofing felt) (brown) | Container Corp. Santa Clara (clay-coated box board) (brown) | San Jose/ Santa Clara Water Pollution Control Plant* | Kimberly-Clark, Fullerton (tissue) (white) | Simpson Ponoma (fine paper) | Pomona Water Reclamation Plant† |
|---|---|---|---|---|---|---|
| Total Solids | 1,000 | — | — | 700 | — | — |
| Volatile Solids | — | — | — | 10 | — | — |
| Fixed Solids | 30 | — | — | 10 | — | — |
| Suspended Solids | 40 | — | — | 30 | — | <65 |
| Dissolved Solids | 1,000 | 1,000 | 1,011 | 650 | <500 | 555 |
| pH | 6-8 | 7 | 7.5 | 8-10 | 7.0-7.5 | 7.1 |
| Hardness | 14 grains | 100 ppm | 302 | 300 | <250 | — |
| Manganese | Tr | 0.5 | — | 30 | — | — |
| Chloride | Tr | 260 | 3.70 | 90 | <100 | — |
| Silica | Tr | — | 26.5 | 9 | — | — |
| Alkalinity | (see pH) | — | 251 | 125 | — | — |
| Aluminum | Tr | 10 | — | .05 | — | — |
| Sulfate | Tr | — | 173 | 300 | <100 | — |
| Turbidity‡ | light | JTU <10 | — | 1.3 mg/L | — | 1.3 NTU |
| Conductivity | — | — | — | 1,100 | <800 | — |
| Color | immaterial | — | 1.9 | — | <10§ | <10** |
| Iron | Tr | 5.0 | — | 0.2 | <1.0 | — |
| Temperature | 130 F mat. | 80°F | — | 15°C (54°F) | <27°C (<80°F) | 22°C (72°F) |

*Source:* Office of Water Recycling. Water Recycling in the Pulp and Paper Industry in California, 1982 (Requirements from responses to CDMJOWR survey, Winter 1980).

* Effluent quality from Plant records.

† Plant records, December 1979.

‡ NTU = Nephalometric turbidity unit; JTU = Jackson turbidity unit, other values as noted.

§ As received (not filtered).

** Filtrate from 0.45 mu filter.

## Industrial Cooling and Boiler Systems

Within different industries, different systems for cooling and boiling are used, with power plants generally using one major heat exchanger (condenser), and industries such as refineries and petrochemical manufactures operating a series of exchangers of different sizes, materials, heat loads, and accessibility. Boilers and cooling systems are usually self-contained and separate from the manufacturing processes, and thus typically have a different standard of water quality from that used within the rest of a manufacturing or utility plant.

Direct contact cooling water is either sprayed or cascaded over the material to be cooled, or alternatively, smaller materials may be dipped into cooling tanks. In open recirculating cooling systems, heat is allowed to dissipate from cooling water so the water can be re-used in the cooling process again. Heat is dissipated primarily through evaporation, which accounts for about two percent of water lost from a system. Drift (water lost to wind and overspill) accounts for a very small percentage (about 0.005%, depending on design) of water lost from a cooling system. The greatest amount of water loss is due to blowdown. Blowdown occurs when water high in concentrated salts in the cooling tower (post-evaporation) is bled off and replaced with low-salt makeup water. These processes can reuse water three to ten times before the increased risk of scaling and suspended solids becomes too high to allow effective operation of the systems.

Cooling towers have the potential to develop four general water quality problems: scaling, corrosion (primarily pitting and crevice), biological growths (can induce corrosion) and fouling. Calcium and magnesium salts can cause scaling, with the silicate salts a particular problem. Metals can cause corrosion; high total dissolved solids and total suspended solids are good indicators of corrosion potential, along with high fluctuations of the temperature and pH. Ammonia, suspended solids, and phosphorus can all serve as nutrient stimulation for biological growths and should be reduced. Biological growths are caused by microorganisms, and fouling can be caused by these as well as debris and suspended solids in the water. The temperature of the water must also be low enough to be effectively used as a coolant.

For recirculating cooling purposes and cooling tower makeup water, reclaimed water quality is typically acceptable for up to 5 cycles. Pre-treatment of reclaimed water is usually required, frequently with a lime classification process, before it is put into a recirculation system. Frequently this is still less expensive than the cost of many other sources of water. Scale inhibitors are commonly used in the cooling water industry, which impact blowdown and ultimate disposal options for the waste streams. Table 3.5 shows the recommended levels of various parameters in water used for cooling towers.

Boiler-feed water quality requirements are dependent upon the pressure at which a boiler is operated, as shown in Table 3.6. The higher the pressure, the higher the water quality required. Very high pressure boilers require makeup water of distilled or desalted quality, and untreated reclaimed water would not be suitable for operating these systems. This may not be an excessive cost if a system is already set up to allow for pre-treatment.

The hardness of water used in boiler water make-up must be reduced to close to zero to make the application feasible. Like cooling towers, calcium and magnesium salts and silicates, as well as aluminum, must be reduced to prevent scaling. Alkalinity must be controlled in order to avoid foaming and deposits on the boiler feed system, and bicarbonate alkalinity in particular must be avoided so as to avert the release of carbon dioxide, which can corrode steam-using equipment.

**Table 3.5**
**Recommended cooling water quality criteria for make up water recirculating systems**

| Parameter* | Recommended limit† |
|---|---|
| Chloride | 500 |
| Total dissolved solids | 500 |
| Hardness | 650 |
| Alkalinity | 350 |
| pH‡ | 6.9-9.0 |
| Chemical oxygen demand | 75 |
| Total suspended solids | 100 |
| Turbidity (NTU)‡ | 50 |
| Biological Oxygen Demand‡ | 25 |
| Organics§ | 1.0 |
| Ammonia as nitrogen‡ | 1.0 |
| Phosphate‡ | 4 |
| Silica Dioxide | 50 |
| Aluminum | 0.1 |
| Iron | 0.5 |
| Manganese | 0.5 |
| Calcium | 50 |
| Magnesium | 0.5 |
| Bicarbonate | 24 |
| Sulfate | 200 |

*Source:* EPA Guidelines for Water Reuse 1992.
* All values in mg/L except pH.
† Water Pollution Control Federation, 1989.
‡ From Goldstein, Wei, and Hickes. 1979.
§ Methylene blue active substances.

## Constructed Wetland and Aquatic Plant Systems

Applications of treated wastewater to wetland and aquatic pond systems must be free of unreasonable risks to aquatic life, waterfowl, and public health. In considering the applications of treated wastewaters to wetlands, the relationship between hydrology and the ecosystem characteristics must also be recognized; factors such as velocity, flow rate, renewal rate, and frequency of application will all influence the reclaimed water quality which can be accommodated by a natural or constructed wetlands system. Wetlands also have the potential for

**Table 3.6**
**Recommended industrial boiler-feed water quality criteria**

| Parameter* | Low pressure (<150) | Intermediate pressure (150-700 psig) | High pressure (>700 psig) |
|---|---|---|---|
| Silica | 30 | 10 | 0.7 |
| Aluminum | 5 | 0.1 | 0.01 |
| Iron | 1 | 0.3 | 0.05 |
| Manganese | 0.3 | 0.1 | 0.01 |
| Calcium | —† | 0.4 | 0.01 |
| Magnesium | —† | 0.25 | 0.01 |
| Ammonia | 0.1 | 0.1 | 0.1 |
| Bicarbonate | 170 | 120 | 48 |
| Sulfate | —† | —† | —† |
| Chloride | —† | —† | —† |
| Dissolved solids | 700 | 500 | 200 |
| Copper | 0.5 | 0.05 | 0.05 |
| Zinc† | 0.01 | 0.01 | —† |
| Hardness | 350 | 1.0 | 0.07 |
| Alkalinity | 350 | 100 | 40 |
| pH, Units | 7.0–10.0 | 8.2–10.0 | 8.2–9.0 |
| Methylene blue active substances | 1 | 1 | 0.5 |
| Carbon tetrachloride extract | 1 | 1 | 0.5 |
| Chemical oxygen demand | 5 | 5 | 1.0 |
| Hydrogen sulfide | —† | —† | —† |
| Dissolved oxygen | 2.5 | 0.007 | 0.0007 |
| Temperature, °F | —† | —† | —† |
| Suspended Solids | 10 | 5 | 0.5 |

*Source:* USEPA, 1992.
* Recommended limits in mg/L except for pH (units) and temperature (°F).
† Accepted as received (if meeting other limiting values); has never been a problem at concentrations encountered. (Was not measured or tested, but accepted as okay by the lab.)

wastewater treatment and application; use as a treatment system is not addressed here, as the details of treatment lie outside the scope of this report.

From a public health and environmental health viewpoint, natural systems have potentially more points of contact with the environment and with the public, primarily because a larger land area is involved in the system. Discharge monitoring is complicated because indicator organisms are diffuse. Careful design and application of reclaimed water must take into account the $BOD_5$ loading of a wetland area, and its loading capacity; the best water depth for a given system; the detention time of the basin; and physical design factors such as flow monitoring and control, outlet structure control and vector control necessary to prevent the spread of disease.

Aquatic systems (which contain floating plants or submerged aquatic plants) also have a high potential for use of reclaimed water, but the volume of water they can accommodate must be considered prior to application. Hydraulic detention time, hydraulic loading rate, and organic loading rate, are the most common parameters used when attempting to size a new aquatic system, and should be checked prior to adding reclaimed water to a system. A constant depth is required for most plants, so the flow balance of the ecosystem should be calculated, as well as the dissolved oxygen (DO) if mosquito fish are to be established to control the insect population. Temperature requirements for the plants in the system may limit the temperature of the reclaimed water that can be added. Average loading of a system without aeration should not exceed 100 kg/ha-d for $BOD_5$, in order to avoid odors and increased mosquito populations.

### Recreational and Aesthetic Impoundments

In man-made water bodies, reclaimed water can serve as a valuable augmentation resource. Aside from the always-important public health considerations, aesthetics must be taken into account when reclaimed water is stored in impoundments, and ammonia, phosphorus and other nutrients may need to be removed in order to control algae blooms and odor.

### Stream Augmentation

Maintenance of stream flows and the enhancement of wildlife habitat can be accomplished with reclaimed water, keeping in mind the designated use of the stream and aesthetics of the area, as mentioned above. A system that is expected to sustain aquatic systems may also require phosphorus and/or nitrogen control, in order to prevent algal blooms. However, a very low level or non-detectable level of chlorine and ammonia would be required in order to keep the toxicity low enough for fish and other aquatic animals. The specific water quality of reclaimed water used for stream augmentation is typically determined during the National Pollution Discharge Elimination System permitting process.

## SUMMARY OF WATER QUANTITY & QUALITY REQUIREMENTS FROM LARGE WATER USERS SURVEY

### Quantity Requirements by Industry

While exact volumes required by industry varied widely, certain general trends were noted after studying the compiled information from the survey database. The data was taken from 32

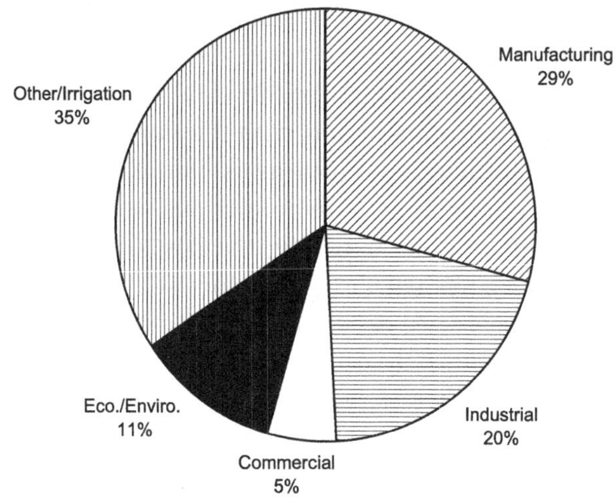


**Figure 3.1 Industry use percentages (by response)**

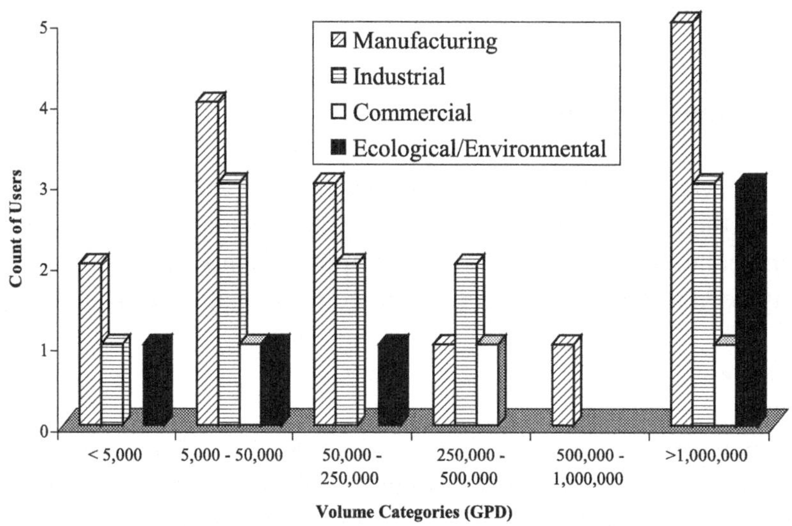


**Figure 3.2 Distribution of volumes across industry survey responses**

different utilities and 53 larger water users, with specific amounts garnered from the utilities and general volumes given by the industries and companies that use reclaimed water.

Of the responses received, approximately 35% of the respondents used the water for irrigation, either of residences, public areas, parks, schools or cemeteries. (See Figure 3.1) It should be noted that the survey was targeted toward large industrial and municipal users recyclers, not irrigation interests. The second largest group responding was the manufacturing industries, with 29%, which largely used the water in industrial processes, with a smaller percentage actually used in the manufacturing processes themselves. While those using reclaimed water for irrigation and commercial uses made up only 40% of the responses to the survey, the number of customers served in those areas made up an overwhelming 96% (See Figure 3.2). This was balanced by the

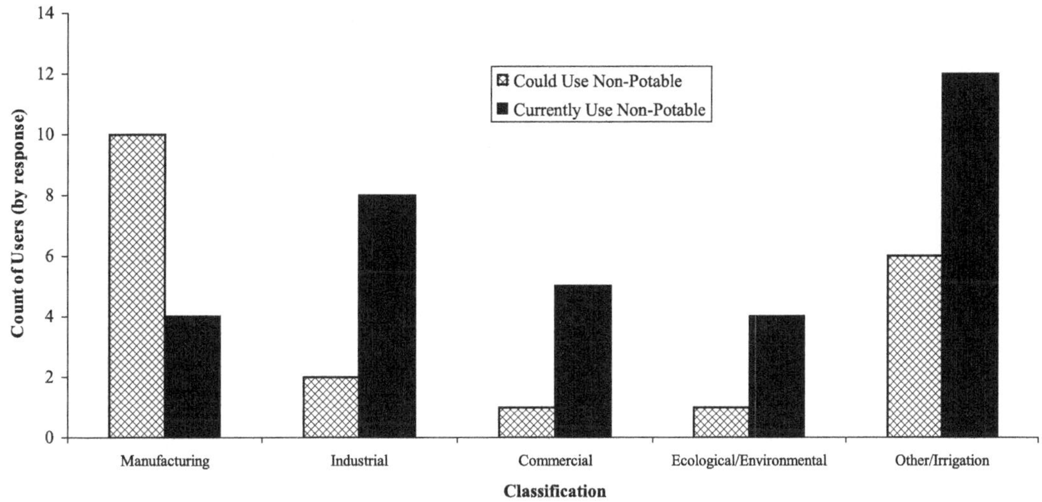


**Figure 3.3 Count of potential vs. current reclaimed water users (by response)**

fact that the large-scale operations made up the largest percentage of water users surveyed which used more than 1 million gallons per day. The number of providers of irrigation water outnumbered these significantly in most of the smaller-volume operations categories. In essence, the amount of reclaimed water purchased for use in manufacturing processes and by irrigation operations was potentially the same, but the demand for industry was significantly larger per individual user than that required by irrigation operations. A consistently available volume of reclaimed water would thus be more critical for a large industrial customer.

Note that, although this report focuses on non-irrigation use, the following figure and tables show irrigation use so as to show complete data from the surveys.

The two areas with the largest potential demand by survey respondents for reclaimed water were manufacturing and irrigation, with a potential increase within the manufacturing sector of 70%, and a potential increase of 33% for irrigation users, according to the responses received from the survey. (See Figure 3.3). Again, keeping in mind the larger volumes used by manufacturing users, it appears that there is a significant demand for a consistent quality and quantity of reclaimed water.

Ecological/Environmental users among those surveyed had the smallest demand for reclaimed water. This was in part because the application of reclaimed water to groundwater, wetlands and aquifers was usually not the most cost-effective application. Ecological/ Environmental use of reclaimed water has frequently been implemented to resolve issues such as the additional cost of other disposal measures when there was no demand for reclaimed water, or water rights and water quantity demand dictated. There were also typically stricter standards in the state regulations for use of reclaimed water in ecological applications than there were in industrial and irrigation applications. The volume of commercial use was the most difficult to quantify as to the accuracy of the data represented here; many commercial outlets get their water from the same multi-user systems that residential and municipal irrigation consumers do, so the number of commercial users and the volume they used could be greater than recorded here.

All of the applications examined in this survey had an overwhelmingly high percentage of year-round reclaimed water use (as was intended under the guidelines for this report). The period

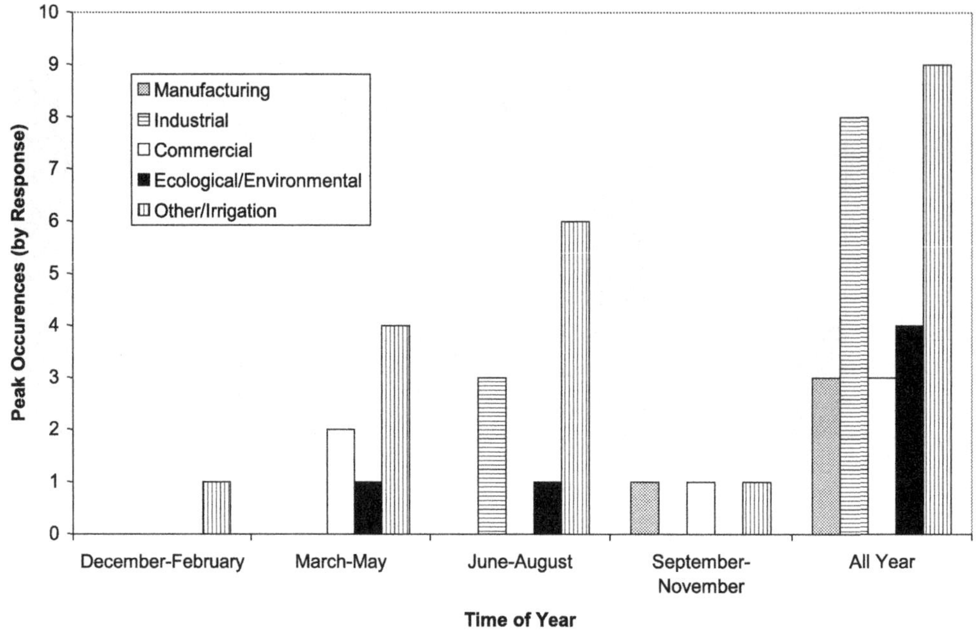


**Figure 3.4 Peak use during the year by industry (by response)**

of June through August had the greatest number of respondents stating that it was their peak usage period, with the greatest volume of reclaimed water distributed. Irrigation applicants made up the largest percentage of users, and industrial cooling towers made up the second largest percentage, with ecological/environmental use the third. March through May was the second highest peak time, with the percentages of cross-industry use holding true for this time of year as well. Peak use volumes based on the survey results are typically at a ratio of about 2:1 to 1.5:1, in comparing the maximum to the daily average. (See Figure 3.4).

A consistent time for peak use over the course of an average day was also less common than operations where a 24-hour steady demand was the norm. Of the respondents from various utilities, over 70% stated that their operation did not have any discernible peak demand period. For those that did, it was usually between the hours of 12 am to 4 am, or 4 am and 8 am, and usually for either irrigation purposes or industrial/commercial cooling towers (See Figure 3.5). This demand is understandable because many irrigators prefer watering at night, when use is low, and because time use restrictions have been enacted for reclaimed water use for irrigation in some entities in order to protect the public from exposure. It is also the most efficient time for watering to take place, since it allows irrigation water to trickle into the ground and soak in before the daylight hours, and thus avoids application when temperatures are high enough to evaporate the moisture before it would have a chance to penetrate.

## ADDITIONAL SURVEYED WATER QUALITY REQUIREMENTS BY INDUSTRY, BEYOND STATE REGULATIONS

While many reclaimed water consumers were satisfied with the effluent water quality provided by reclaimed water utility facilities, respondents of the industry and municipal surveys mentioned certain other requirements. Given the way the survey was constructed, a range of

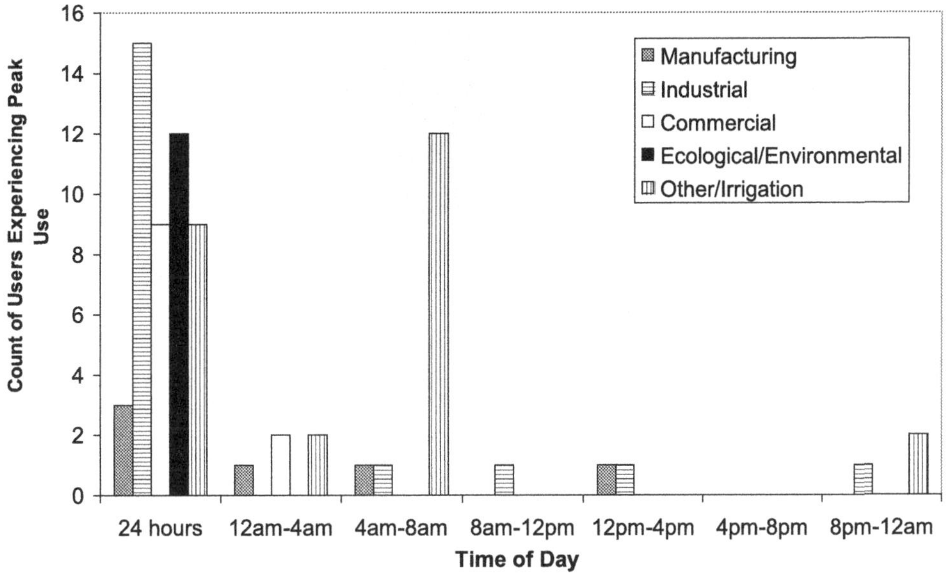


**Figure 3.5 Peak use for time of day**

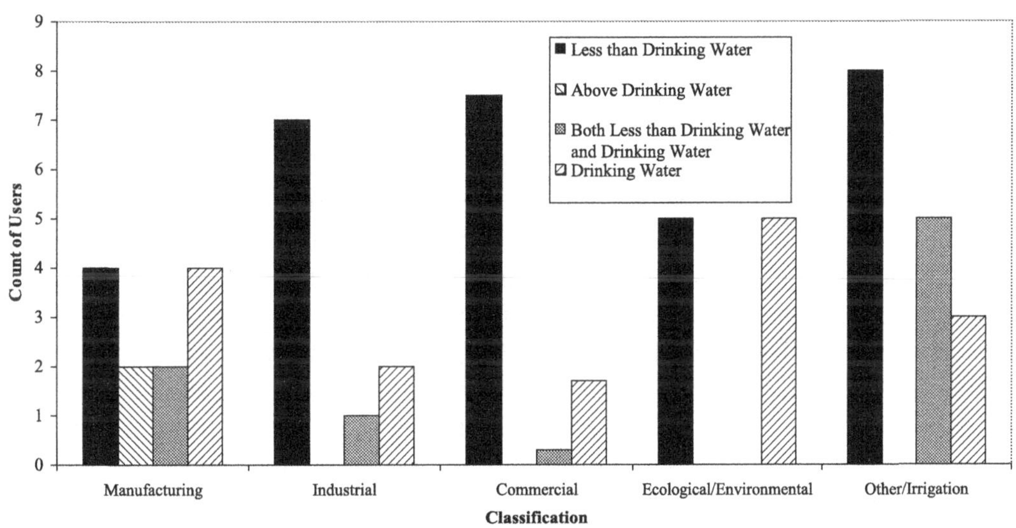


**Figure 3.6 Water quality standards by industry (by response)**

responses was inevitable. Many answered the questions regarding the water quality required with more than one answer. One use would require water conforming to drinking water standards for one process and not for another process. Trying to separate the answers into percentages of respondents who required "better than drinking water quality" reclaimed water, versus those who needed to "meet drinking water quality" standards, was sometimes difficult. A comparison by classification of user is shown in Figure 3.6.

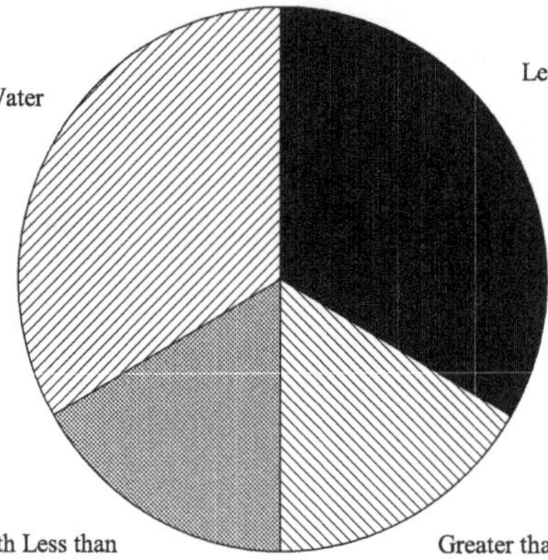


**Figure 3.7 Manufacturing water quality requirement percentages**

### Manufacturers – Paper, Carpet, Dyes, and Petroleum

Consistency of water quality as well as low color, low levels of chlorine, ammonia, phosphates, and low pH and metals concentrations, were cited as the primary water quality parameters required in industries that used reclaimed water. Approximately 33% to 50% of those responding said that their processes needed water that met drinking water standards, 17% stated that they needed water that exceeded them, with a third to a half of those surveyed requiring water that did not necessarily meet drinking water standards. (See Figure 3.7). The majority did not apply reclaimed water in the critical portions of their manufacturing process, preferring potable water in order to be assured of no chemical interference in producing their materials. A. L. Johnson Co. Foundry, which manufactures aluminum castings, stated they must use potable water in their manufacture of the basic molds in which the aluminum casts are made. There were, however, exceptions. The Total Petroleum oil refinery applies small volumes of reclaimed water in their desalting of crude oil, since it is subject to further processing later. The Chevron Oil Refinery in East Bay, California, used reclaimed water, but only after a list of specific constituent limits had been met by their municipal supplier , including low free residual chlorine, low phosphates and ammonia, low Total Organic Carbon (TOC) and a Rainbow Trout acute bioassay (as part of the discharge requirements).

Reclaimed water is often used in the 'secondary' processes, where color and consistency is not as vital, such as rinsing and intermediate washings. The user typically treats reclaimed water before it enters any process where it might affect the color or texture of the final product, often being softened and filtered before application. A consistent level of water quality is essential at this point, since once the level of treatment necessary is determined, the formula for the application of treatment chemicals is usually automated. A change in water quality, good or bad, can throw off treatment by the customer and ruin an entire run of paper or carpet. The survey respondents generally did not state desired or required levels of specific contaminants. Stated levels for chlorine were around 1 mg/L. Stated levels of hardness (as calcium carbonate) were variable with a range from 150 mg/L to 450 mg/L. Both were lower than the recommended levels for cooling towers (500 mg/L and 650 mg/L), and in some cases were below that recommended

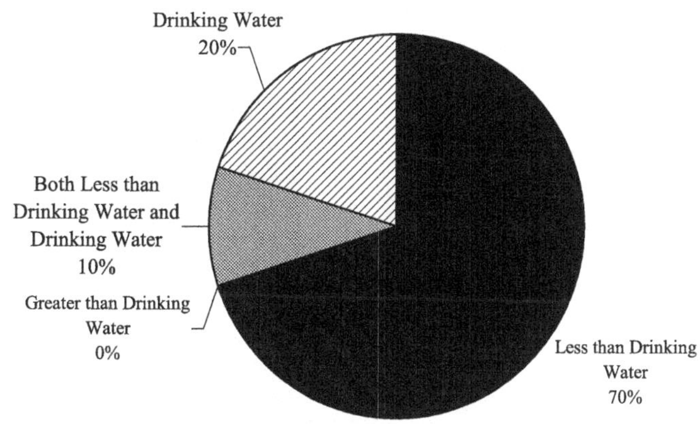


**Figure 3.8 Industrial water quality standard requirements**

by the EPA for the industries surveyed. Waters near neutral were desired. Confidence in the non-potable water provider's ability to provide a consistent level of water quality, and provide ample warning in case of accidents at a treatment facility, was described as the main concern of manufacturers using reclaimed water.

## Industrial Use – Cooling & Boiling Towers

The softer and more disinfected the water, the better, according to respondents to this survey. While several respondents cited softness as a major issue, the need for a high level of disinfection and low levels of nutrients, particularly ammonia, phosphates and nitrates, which serve as food for bacteria and pathogens, was a much more strongly expressed concern. Approximately 20% of those surveyed stated that they needed water which met drinking water standards, with about 70-75% requiring water which did not necessarily meet drinking water standards. No industrial consumers stated that they required water that they purchased to exceed drinking water standards. (see Figure 3.8)

Most mentioned that effluent water and disinfected secondary effluent were both usable in cooling towers, but that a certain amount of pre-treatment to add biocides and soften the water was necessary when secondary effluent was used in recirculating cooling systems. Boilers require water quality comparable to that of distilled water, and thus any reclaimed water which was used in boilers was subjected to extensive pre-treatment, and was only purchased if there was difficulty in obtaining potable water, if the local water supply did not have a hardness problem or if a treatment system was already in place and prepared to handle additional disinfection and softening.

The petroleum industry, for instance (according to the representative from the Conoco Oil refinery) uses potable water in their boilers, which has been softened with hot lime for silica removal, followed by sodium cation exchange. One of the more unique users was a customer of the Tampa Wastewater Treatment Plant, an anhydrous ammonia storage facility, which produces power from refuse products. They were able to use reclaimed water in their cooling systems, and also recycle the heat transferred to the reclaimed water through the facility in order to complete the process.

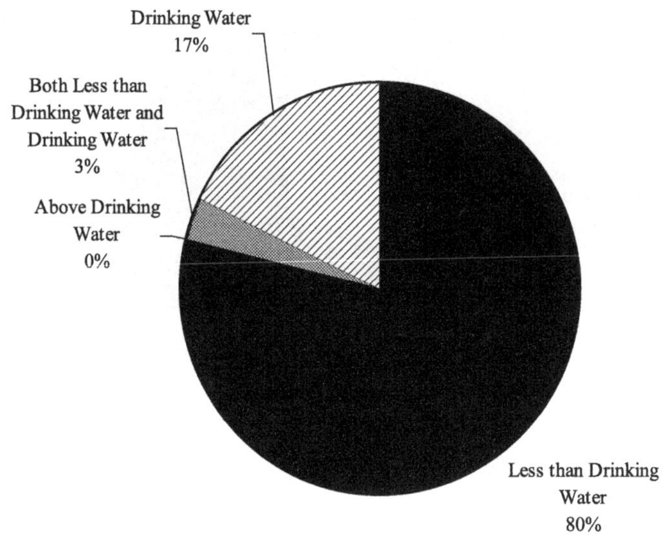


**Figure 3.9 Commercial water quality standards**

### Commercial Use – Irrigation, Car Washes, Domestic (Inside) Lines

While the responses from the commercial sector made up the smallest percentage of this survey, a few concerns unique to this particular use were expressed. Approximately 17% said that they needed water that met drinking water standards, with 70-80% requiring water that did not necessarily meet drinking water standards. No commercial consumers stated that they required water to exceed drinking water standards. (See Figure 3.9). Reclaimed water applied in public areas, such as flush toilets and public washes, needed to be aesthetically pleasing, with low color and turbidity as well as being free of pathogens. Also, there are instances around the country where reclaimed water has been used in car washes. Since most agreed that the water supplied by their local reclaimed utility was sufficiently clean to meet their needs, and none of the commercial operations had the ability to apply additional treatment to any reclaimed water purchased, it can be inferred that the state standards for commercial applications are meeting the quality demands of the consumers at this time.

### Ecological/Environmental Use – Wetlands & Groundwater Make-up

Aside from an expressed preference for high-level disinfection for use in ecological and environmental applications, consumers applying reclaimed water to wetlands and groundwater make-up systems emphasized the need for low phosphorus and nitrogen levels as well as low chlorine or dechlorinated water. Avoiding algal blooms and maintaining the low chlorine levels to prevent harm to any wildlife in the ecosystem were the most common reasons cited. It can be assumed that some dechlorination is required for reclaimed water that has been disinfected using chlorine. The Denver Water Department stated that the Washington Park and City Park lakes, and the Denver Zoo bird habitat all inquired about low levels of phosphorus, ammonia, and chlorine in the proposed reclaimed water. Approximately 50% of respondents claimed that they needed water that met drinking water standards, with about 50% requiring water that was less than drinking

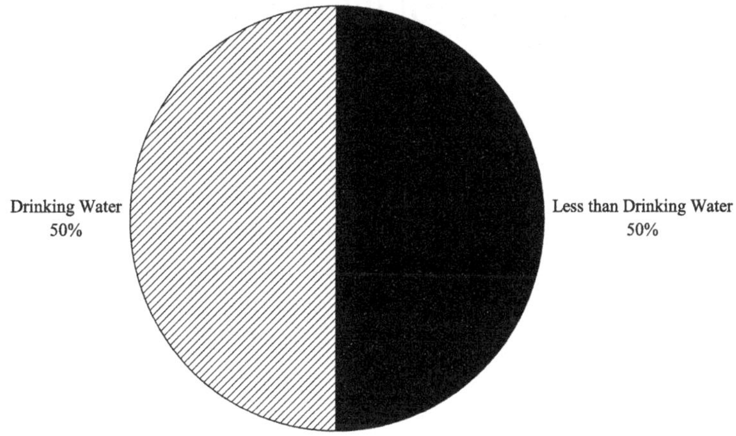


**Figure 3.10 Ecological/environmental water quality standard requirements**

water standard. No ecological or environmental consumers stated that they required water to exceed drinking water standards. (See Figure 3.10) Additionally, the West Basin Municipal Water District reported that they were able to send excess reclaimed water that had been through a micro-fitted reverse osmosis process (as well as tertiary treatment) to the West Coast Barrier aquifer as additional groundwater recharge.

## SUMMARY OF WATER QUANTITY & QUALITY REQUIREMENTS FROM UTILITY SURVEYS (CUSTOMER REQUIREMENTS)

From the results of the survey, it appears that few utilities receive specific water quality requests from their customers beyond compliance with the state standards. Most consumers appear to rely on the state-required levels of disinfection and filtration, and are aware of which level of water quality they wish to purchase, without demanding further treatment from their suppliers. The water quality requirements expressed in the Large User Survey (above) were reiterated in the municipal and utilities surveys, but no requests for further treatment were reported by the utilities surveyed. According to the results of this survey, 20-25% of reclaimed water plants, wastewater treatment plants, and utilities across the country obtain their water from more than one source, and of those, approximately half are from non-potable sources such as stormwater runoff, graywater sources (in-house waste water, from sources other than toilets), and wastewater streams. Approximately a third to half of the utilities surveyed mentioned projected upgrades and planned expansions to their facilities, in order to meet a growing demand for reclaimed water. This was especially true in Florida, where water shortages are an ongoing critical concern, and the need to meet the established discharge requirements has made additional treatment necessary.

In defining potential large users and developing a reclaimed water use program with them, four specific steps should be followed:

Define each potential source water's quality.
Define users' required water quality.
Evaluate treatment needs (if any) for the various source waters.
Select the most appropriate source water.

## SUMMARY OF WATER QUALITY REQUIREMENTS FROM STATE PERMITTING AGENCIES

Water quality requirements for reclaimed water vary from state to state, not only in the numeric standards implemented, but also in the parameters chosen to regulate reclaimed water quality. The main purpose of these standards is the protection of public health, with environmental concerns often addressed separately. The classification of the specific uses and warning levels which constitute 'restricted' use and 'unrestricted use' (and thus contain more stringent standards for protection of public health) fluctuate without apparent consistent correlation, and the level of protection and monitoring required in order to protect the environment also varies with climatic conditions and use. The following summary of the regulatory limits examines the basis (either explicit in the regulations, or derived from the literature) for some of the states' regulatory limits on recycling/reuse of wastewater, as well as the differing definitions of treatment, industry application and public safety.

### Access: Restricted, Unrestricted, and Levels of Control on Reclaimed Water Applications

Most states have established their numeric regulatory standards based upon the location where the reclaimed water is to be used and the level of public access to the area, surrounding land uses or processes, the nearest location of potable water sources and the location and types of warning signs to be posted. Protection of both public health and the environment is considered when setting compliance standards, but the inferred health risks to the general population are given precedence. While the numeric standards for reclaimed water systems with unrestricted public access, and the standards for restricted access reclaimed water applications, are comparable across states, there are no hard-and-fast rules which any utility or reclaimed water provider can apply for applications in different geographic areas, such as can be applied to drinking water or the minimums for wastewater treatment. Many states have few references to industrial uses in their regulations. Table 3.7 provides a summary of industrial and impoundment reclaimed water regulations. Table D.2 provides comparisons of the access levels and allowed applications for different levels of treatment for all uses, including irrigation. Table 3.8 lists corresponding numeric standards. Note that on both tables, the terms reuse, recycled, and reclaimed are used as listed in the regulations.

California, Florida, Nevada, and Texas have detailed regulations regarding both the industries and conditions under which different levels of disinfection, filtration and treatment are applicable for reclaimed water, as well as application of reclaimed water in underground wells and wetlands. However, Arizona, Colorado, Oregon and Utah confine use to simpler definitions of restricted (wherein the water is unlikely to come in contact with members of the public because of distances, hours of application, or physical barriers) or unrestricted (which usually requires warning signs, but few other impediments to exposure for members of the general public). The states with the more detailed regulations generally have more industrial, commercial and ecological use than those without

It can be assumed that as more reclaimed water programs are put into effect, additional standards for different industrial uses will be adopted.

States with more specific regulation levels also usually have more specific environmental concerns. Arizona's legislation includes specific values for wetlands application, with monitoring frequency dependent on rate of flow. Texas makes the distinction between reclaimed water that is

**Table 3.7**
**Access and allowed use across different states**

| State, access level | Definition of access, allowed uses, and setbacks |
|---|---|
| Arizona (restricted) | Industrial use to be considered on an individual basis, but use in cooling systems is allowed. |
| California (undisinfected secondary recycled water) | Recycled water used for flushing sanitary sewers. |
| California (disinfected secondary-2.2 recycled water) | Supply for restricted recreational impoundments and for any publicly accessible impoundments at fish hatcheries. |
| California (disinfected secondary-23 recycled water) | Recycled water for industrial or commercial cooling or air conditioning that does not involve the use of a cooling tower, evaporative condenser, spraying, or any mechanism that creates a mist. |
| | Recycled water may be used in:(1) Industrial boiler feed, (2) Nonstructural fire fighting, (3) Backfill consolidation around non-potable piping, (4) Soil compaction, (5) Mixing concrete, (6) Dust control on roads and streets, (7) Cleaning roads, sidewalks and outdoor work areas and (8) Industrial process water that will not come into contact with workers. |
| California (disinfected tertiary recycled water) | Non-restricted recreational impoundments (subject to conventional treatment). |
| | Non-restricted recreational impoundments without conventional treatment, provided the recycled water is monitored for the presence of pathogenic organisms in accordance with section 60305.1 (*Giardia*, enteric viruses, and *Cryptosporidium* screening standards). |
| | Recycled water used for industrial or commercial cooling or air conditioning that involves the use of a cooling tower, evaporative condenser, spraying or any mechanism that creates a mist. Provided a drift eliminator shall be used whenever the cooling system is in operation, and a chlorine, or other, biocide shall be used to treat the cooling system recirculating water to minimize the growth of *Legionella* and other microorganisms.) |
| | Recycled water can be used for the following: (1) Flushing toilets and urinals, (2) Priming drain traps, (3) Industrial process water that may come into contact with workers, (4) Structural fire fighting, (5) Decorative fountains, (6) Commercial laundries, (7) Consolidation of backfill around potable water pipelines, (8) Artificial snow making for commercial outdoor use, and (9) Commercial car washes, including hand washes if the recycled water is not heated, where the general public is excluded from the washing process. |

(continued)

**Table 3.7 (Continued)**

| State, access level | Definition of access, allowed uses, and setbacks |
|---|---|
| Florida (secondary treatment & basic disinfection) | Industrial uses permitted. Reclaimed water may be used for open cooling towers, if a 300' setback is maintained from the property line.<br>Reclaimed water may be used for toilet flushing in commercial or industrial facilities or buildings. Reclaimed water may be used for toilet flushing in motels, hotels, apartment buildings, and condominiums where the individual guests or residents do not have access to the plumbing system for repairs or modifications. Reclaimed water shall not be used for toilet flushing in any residential property or dwelling unit where the residents have access to the plumbing system for repairs or modifications.<br>Reclaimed water may be used to provide water for fire protection.<br>Reclaimed water may be supplied to fire hydrants. Hydrants supplied with reclaimed water shall be color-coded and shall have no connection to the potable water supply.<br>Reclaimed water may be used to provide water for fire protection in sprinkler systems located in commercial or industrial facilities or buildings. Reclaimed water may be used to provide water for fire protection in sprinkler systems located in motels, hotels, apartment buildings, and condominiums where the individual guests or residents do not have access to the plumbing system for repairs or modifications. Such sprinkler systems shall be color-coded and shall be supplied only by reclaimed water.<br>Reclaimed water may be used for dust control at construction sites.<br>Reclaimed water may be used for indoor or outdoor aesthetic purposes. Examples of aesthetic uses of reclaimed water include decorative pools, fountains, ponds, lagoons, and other aesthetic features.<br>Additional uses that are approved include: (a) Water supply for commercial laundries. (b) Vehicle washing. (c) Flushing of sanitary sewers and reclaimed water lines. (d) Mixing of concrete. (e) Manufacture of ice for ice rinks. (f) Cleaning roads, sidewalks, and outdoor work areas. |
| Florida (secondary treatment, filtration & high-level disinfection) | Open cooling towers with no setback.<br>Wetlands creation, restoration, or enhancement. Public access to the wetlands shall be restricted. |
| Nevada (B) – limited access | An impoundment where all activities are prohibited and human contact with the treated effluent does not occur. |
| Nevada (C) – unrestricted | Impoundments where full body contact with the treated effluent cannot reasonably be expected.<br>No setback from domestic water supplies. |
| Oregon (IV – unrestricted) | A body of reclaimed water in which no limitations are imposed on body-contact water recreation activities. |
| Oregon (III – restricted) | Recreation is limited to fishing, boating, and other non-body-contact water recreation activities. |

(continued)

**Table 3.7 (Continued)**

| State, access level | Definition of access, allowed uses, and setbacks |
|---|---|
| | Use of reclaimed water in evaporative cooling systems shall be approved only if the user can demonstrate that aerosols will not present a hazard to public health. |
| | Construction/industrial use. |
| Oregon (II – restricted) | Same as Oregon III, but samples are collected weekly, instead of tri-weekly. May not be used in an impoundment. |
| Oregon (I – prevented) | Construction/industrial use. |
| Texas (general) | Reclaimed water shall not be utilized in a way that degrades ground water quality to a degree adversely affecting its actual or potential uses. |
| Texas (I – unrestricted) | Fire protection, either in internal sprinkler systems or external fire hydrants; maintenance of impoundments or natural water bodies where recreational activities, such as wading or fishing, are anticipated even though the water body was not specifically designed for such a use; toilet or urinal flush water; other similar activities where the potential for unintentional human exposure may occur. |
| Texas (II – restricted – non-pond) | Maintenance of impoundments or natural water bodies where direct human contact is not likely. |
| | Soil compaction or dust control in construction areas where application procedures minimize aerosol drift to public areas. |
| | Cooling tower makeup water. Use for cooling towers, which produce significant aerosols adjacent to public access areas, may have special requirements. |
| | Irrigation or other non-potable uses of reclaimed water at a wastewater treatment facility. |
| Texas (II – restricted – pond) | See Texas (II-restricted – non-pond ) conditions. |
| Utah (I – human exposure) | Urban uses, which include toilet flushing, fire protection, and other uses with similar potential for human exposure. |
| | 500 ft. setbacks for impoundment, away from potable sources. |
| Utah (II – no human exposure) | Impoundments of wastewater where direct human contact is not allowed or is unlikely to occur. |
| | Cooling water. Use for cooling towers, which produce aerosols in populated areas, may have special restrictions imposed. |
| | Soil compaction or dust control in construction areas. |
| | 500 ft. setbacks for impoundment, away from potable sources. |

**Table 3.8**
**Summary of numeric levels and parameters across different states**
**(All regulations are final, in-place rules or legislation, unless otherwise noted.)**

| State (level of access) (status of regulation) | Fecal coliform (org/100 ml) | *E. Coli* (org/100 ml) | $BOD_5$ (mg/L) | TSS (mg/L) | Turbidity (NTU) | Chlorine residual (mg/L) | pH | Other |
|---|---|---|---|---|---|---|---|---|
| Arizona (restricted) | 200 av./1000 max. | N/A | N/A | N/A | 1 | N/A | 4.5-9 | N/A |
| Arizona (wetlands marshes) | 1000 gm/4000 max. | N/A | N/A | N/A | N/A | N/A | 6.5-8.6 | 0.5 pH change unit/day<br>6 D.O.<br>Temp. not to interfere with aquatic life |
| California (undisinfected secondary recycled water) | N/A | N/A | N/A | N/A | N/A | N/A | N/A | Organic matter is stabilized, non-putrescible, and contains D.O. |
| California (disinfected secondary-2.2 recycled water) | 2.2 MPN, over 7 days; <23 max. | N/A | N/A | N/A | <2 in 24 hours; <5 for 5% in 24 hours; 10 max | N/A | N/A | N/A |
| California (disinfected secondary-23 recycled water) | 23 MPN, over 7 days; <240 max. | N/A | N/A | N/A | <2 in 24 hours; <5 for 5% in 24 hours; 10 max | N/A | N/A | N/A |

(continued)

**Table 3.8 (Continued)**

| State (level of access) (status of regulation) | Fecal coliform (org/100 ml) | *E. Coli* (org/100 ml) | $BOD_5$ (mg/L) | TSS (mg/L) | Turbidity (NTU) | Chlorine residual (mg/L) | pH | Other |
|---|---|---|---|---|---|---|---|---|
| California (disinfected tertiary recycled water) | 2.2 MPN med. for seven days; <23 MPN for 30-day period; 240 max. | N/A | N/A | N/A | <2 in 24 hours; <5 for 5% in 24 hours; 10 max. | CT > 450 mg/L-min (>90 minutes) | N/A | MS2 and Polio virus- 3-log removal |
| Colorado (unrestricted) | N/A | 126 ave./235 max. | N/A | | 3 ave./5, 5% max. | N/A | N/A | N/A |
| Colorado (restricted) | N/A | 126 ave./235 max. | N/A | 30 | N/A | N/A | N/A | N/A |
| Florida (secondary treatment and basic disinfection) | 200 avg./800 max. | N/A | <20 $CBOD_5$ | <20 | N/A | 0.5 | 6.0-8.5 | N/A |
| Florida (secondary treatment, filtration, high-level disinfection) | 75% BDL/25 max. | N/A | <20 $CBOD_5$ | 5 | * | 1.0* | 6.0-8.5 | N/A |
| Hawaii (treatment works) | <23 med./240 max. | N/A | 30 ave./ 60 max. | 30 ave./ 60 max. | N/A | 0.1 (subsurface disposal system, non-soil absorption) | N/A | N/A |
| Maryland (DRAFT) | <3 | N/A | <10 | <10 | N/A | N/A | N/A | N/A |
| Nevada (B) | 23 gm/240 max. | N/A | 30 | N/A | N/A | N/A | N/A | N/A |

(continued)

**Table 3.8 (Continued)**

| State (level of access) (status of regulation) | Fecal coliform (org/100 ml) | *E. Coli* (org/100 ml) | $BOD_5$ (mg/L) | TSS (mg/L) | Turbidity (NTU) | Chlorine residual (mg/L) | pH | Other |
|---|---|---|---|---|---|---|---|---|
| Nevada (C) | 2.2 gm/23 max. | N/A | 30 | N/A | N/A | N/A | N/A | N/A |
| Oregon (IV – unrestricted) | 2.2 ave./23 max. | N/A | 30 | N/A | 2 ave./ 5 max. | N/A | N/A | N/A |
| Oregon (III – restricted) | 2.2 ave./240 max. (3 samples per week) | N/A | N/A | N/A | N/A | N/A | N/A | N/A |
| Oregon (III – restricted) | 23 ave./240 max. for 2 con. samples (weekly) | N/A | N/A | N/A | N/A | N/A | N/A | N/A |
| Oregon (I – prevented) | N/A | N/A | N/A | N/A | N/A | N/A | N/A | N/A |
| Texas (I – unrestricted) | 20 ave./75 max. | N/A | 5 | N/A | 3 | N/A | N/A | N/A |
| Texas (II – restricted – non-pond) | 200 ave./800 max. | N/A | 20 (15 for $CBOD_5$ ) | N/A | N/A | N/A | N/A | N/A |
| Texas (II – restricted – pond) | 200 ave./800 max. | N/A | 30 | N/A | N/A | N/A | N/A | N/A |
| Utah (I – human exposure) | BDL ave., <14 max. | N/A | <25 | <25 | <2 ave., 5 max. | N/A | 6-9 | CT – 30 mg/L-min (>30 minutes) |
| Utah (II – no human exposure) | <200 | N/A | <25 | <25 | N/A | N/A | 6-9 | N/A |

* Turbidity and chlorine are monitored continuously and are used in conjunction with the operating protocol to make continuous judgments on whether reclaimed water being produced is of sufficient quality to be released to the reclaimed water system.

stored in restricted ponds, and other restricted access applications, allowing a slightly higher biological oxygen demand for ponds, where the solids can be assumed to settle out over time. California's groundwater recharge permit conditions are based on the level of treatment provided; effluent quality and quantity; spreading area operations; soil characteristics; hydrogeology; residence time; and distance to withdrawal, and are made on a case-by-case basis. However, all of the surveyed states have set specific setbacks for reclaimed water application and storage away from potable water supplies and resources, most varying between 50 to 150 feet. Maryland's proposed standards do not specify a change in regulatory standard that would depend on the specific use of the reclaimed water or the level of public access, as the rest of the states surveyed do, and it is unknown whether Hawaii's proposed standards will address the differences in applications and access.

Monitoring frequency for water quality fluctuates based on the above factors, as well as the speed of flow and the physical barriers imposed by a treatment system or reuse application set-up. Likewise, requirements for control in treatment plants that produce reclaimed water are similar but not identical across the states, some of which detail the labels and controls which must be in place for operation of a reclaimed water plant, while others simply set compliance standards for operators prior to discharge of water from the plant, and refer to state groundwater and NPDES standards for surface water discharges.

All of the states surveyed reserve the right to set additional levels of environmental and public health protection at the discretion of their regulatory agencies and commissions, based upon the permit applicants' descriptions of the reclaimed process proposed.

**Choice of Monitoring Parameters**

The public health risk of contracting disease from pathogenic microorganisms through exposure to reclaimed domestic water is mitigated by treating wastewater, so as to minimize the number of viable pathogenic microorganisms such as bacteria, viruses and protozoans. Acceptable public health risk is determined based on an absence of acute gastrointestinal disorders (the most likely type of disease manifestation) in those persons casually exposed to reclaimed domestic wastewater as it is used for surface irrigation of landscaping, or in industrial complexes. Bacterial protection is ensured through the imposition of limits on fecal coliform bacteria (which is present in higher concentrations than pathogens and are easily and unambiguously detectable) or *Eschericia coli* (*E. coli*), a potential source for water-borne gastroenteritis, which has a comparable infection dose to most other water-borne bacteria. Treatment facilities must apply disinfection protocols such that these standards are met at the point where their discharge is monitored. Viral and protozoan (meaning specifically enteroviruses, and giardia/cryptosporidia parasites) protection is ensured by the imposition of limits for turbidity or total suspended solids, as appropriate, since most virus detection in an aquatic environment is difficult at best. It is common to monitor surrogate parameters such as these to indicate how well treatment processes are working. Many of these pathogens are capable of being aerosolized and distributed through water spray, and thus while the exposure (due to diffusion and lower volumes ingested) and infectious doses are less than they would be for a drinking water supply, it must still be considered a health risk and be regulated and monitored in order to protect the public. Reducing the biological oxygen demand (BOD) for a wastewater stream reduces the organics, which provide food for microorganisms and pathogens, and is necessary from the standpoint of disinfection, since these can interfere with other water treatment processes.

As stated before, Maryland and Hawaii do not currently have any specific reclaimed water quality requirements, although both were in the process of implementing regulations to address this during the data collection period. The proposed rules for Maryland include standards for fecal coliform counts, 5-day Biological Oxygen Demand ($BOD_5$) measurements, and total suspended solids (TSS), similar to those currently in place in most of the other states surveyed. However, there are exceptions. Colorado is unique in requiring biological testing for *Eschericia coli* (*E. coli*) instead of fecal coliforms; and California's tertiary treatment standards include extra disinfection treatment for viruses and pathogens such as meningitis 2 and the Polio virus.

### *Turbidity and Total Suspended Solids*

A significant safety margin for turbidity and viral transmission through reclaimed water has been chosen by most states. A 2 NTU limit is used for unrestricted reclaimed water in California and Oregon, a 3 NTU limit is used in Colorado and Texas (30-day average in TX, only), a less than 3 NTU limit in Utah, and a 5 NTU limit is used in Tucson, Arizona and is currently being proposed in the Arizona draft reclaimed water regulations. For restricted use reclaimed water, a standard for total suspended solids is commonly applied, but again, the levels vary from state to state.

### *Additional Parameters*

Additional standards for chlorine residuals and pH are applied in Florida and Utah, and are proposed for Arizona. California includes requirements for various viruses and bacteria beyond the standard biological and solid standards. Florida requires sampling for *Cryptosporidium* and *Giardia* when reclaimed water is used in open cooling towers. Again, additional parameters are usually added to the standards in areas where environmental concerns and preservation of an unstable ecosystem are an issue.

## Industrial/Ecological Water Quality Requirements Summary

Quantities of water used vary widely from state to state and while information for California, Florida, and Texas is available, the other states reviewed in this study are not currently keeping track of the exact quantities of reclaimed water applied in their states. However, as stated before, all of the states surveyed are interested in increasing reclaimed water use, and are enacting regulations and legislation to protect the public health and environment in anticipation of greater future use. There is a greater need for recharge, industrial and environmental applications of reclaimed water in states where drought is possible, but no other differences were noted in manufacturing, industrial, commercial or irrigation volumes used across different states.

None of the state reclaimed water quality requirements exceed manufacturing requirements, but many of the standards in California and Florida are regarded as adequate for all other uses by the consumers surveyed, without pre-treatment. Per the surveys, the more limited uses in the other states do not have any consumer-proposed water quality standards that are more stringent than the state-imposed regulations. Environmental and ecological uses require more monitoring and forethought in design, in order to prevent harm to an ecosystem by excessive application of water that would be usable at lower volumes. Many of the states still have a case-by-case basis for water quality, monitoring, and volumes applied for many kinds of industrial and environmental reclaimed water application when they are granting a reclaimed water permit (including Colorado, Hawaii, Nevada, and Oregon).

## COMPARISON TO EPA DRINKING WATER STANDARDS

While the SDWA Primary and Secondary Water Quality Standards (designed as public health standards) are not entirely comparable to the EPA's guidelines for water reuse in different industries (designed for incidental public exposure), it should be noted that SDWA standards are frequently the basis for the development of regulatory standards that are applied towards any reclaimed water that may come in contact with the public. It is not required that reclaimed water meet drinking water standards, since the level of exposure is much less than a lifetime's worth of 2 L/day consumption. Drinking water standards are included in Appendix E for reference.

# CHAPTER 4
# SUMMARY AND CONCLUSIONS

## PROJECT SUMMARY

There is demand for and interest in expanded reclaimed water applications, as found both in the literature and in the surveys. Current levels of reclaimed water quality are appropriate for many uses, and those who wish higher quality (manufacturers) are typically in a position to pre-treat the water at their own facilities. Environmental applications of reclaimed water are many, from wetlands and stream augmentation to ground water make-up to recreational impoundments, but application of reclaimed water in these arenas can be more difficult than other applications due to more stringent water quality, monitoring, and flow requirements. The charts and graphs included in this report cover state regulatory reclaimed water quality standards, currently recommended EPA guidelines, and proposed industrial standards mentioned by the survey respondents.

## APPROACH - SURVEY AND LITERATURE REVIEW

Obtaining complete information for the surveys was complicated by the necessity of contacting some of the respondents directly over the phone. Additionally, several of the survey respondents were unfamiliar with the specific water quality parameters, which their industry operated under, and could only comment on the basic regulatory requirements.

It was difficult to find material in published literature about industrial reclaimed water, and not much was available beyond government studies. But there are EPA-sponsored pilot projects in the works in several states, which, if they were followed up on, might provide some useful data.

## SIGNIFICANT RESULTS

Consistent levels of water quality, and confidence in disinfection levels, were the two most commonly reported concerns from consumers and utilities in this survey. Application is the deciding factor in the standard of reclaimed water quality required. Geographic differences do not affect public health standards, and the percentage of applications that affect the surrounding environment in any unique way is very low; most are regulated on a case-by-case basis in the states where environmental and ecological applications are permitted. Enacting tertiary standards for the states that currently do not have them could be useful in stimulating interest among industries in those states in applying reclaimed water beyond the current applications. No truly consistent reasoning for the minor differences in disinfection and turbidity standards could be found.

## ADDITIONAL RESEARCH NEEDS

More research for a consistent disinfection standard, comparing *E. Coli* to fecal coliform counts, needs to be done. Results from Colorado's use of the *E. Coli* standard should be compared to other states after it has been in place for more than a year. Several other states, including North Carolina, also have reclaimed water programs, and data from their industries should be included in further investigations of water quality standards. Specific justifications for the differing state standards needs to be compiled, so that the utilities in each state can make their own decisions regarding the need for further disinfection or filtration of the water they provide, as well as a comparison of any public health records regarding reclaimed water exposure in those states. As the use of reclaimed water becomes more common, the standards for protection of the public health and the environment should take these expanded applications into account. Also, as more experience with industrial use of reclaimed water is gained, the successes, problems, and failures should be documented and summarized.

# CHAPTER 5
# RECOMMENDATIONS TO UTILITIES

This report is the starting point for planning by utilities for new or expanded operations based on the needs of large users. Investigations of proposed, specific uses should be made of individual industries, manufacturers, and environmental applications for both their specific quantity and quality needs. State regulators should be contacted during the planning phases for details about both areas. Reclaimed water can be a valuable resource in water-scarce areas, or a viable disposal method. The need for dependability in areas where demand fluctuates and may affect the quality of reclaimed water, should be taken into account, as well as the preventive measures necessary to guarantee consistent water quality.

# APPENDIX A
# LIST OF INTERNET DATABASES FOR STATE AGENCIES

**LIST OF INTERNET DATABASES FOR STATE AGENCIES**

Arizona's State Web Page: http://www.sosaz.com/

Arizona's Administrative Code: http://www.sosaz.com/public_services/Table_of_Contents.htm
Title 18, Articles 9 & 11

California Department of Water Resources Home Page: http://www.dwr.water.ca.gov/

California Code of Regulations: http://www.calregs.com/default.htm
Title 22 (Social Security), Division 4, Chapter 3: Water Recycling Criteria

Colorado Department of Public Health and Environment Water Quality Control Commission: Regulation 84 –
http://www.cdphe.state.co.us/op/regs/waterregs/100284.pdf

Florida – Chapter 62-600 and 62-610
http://www.dep.state.fl.us/water/reuse

Hawaii – Department of Health, Title 11
http://www.hawaii.gov/doh/rules/11-62.pdf

Nevada Administrative Code - Chapter 445A, .275-.280
http://www.leg.state.nv.us/NAC/nac-445A.html

Oregon Administrative Rules contain OARs filed through March 15, 2001
DEPARTMENT OF ENVIRONMENTAL QUALITY - DIVISION 55
http://arcweb.sos.state.or.us/rules/OARS_300/OAR_340/340_055.html

Texas Natural Resource Conservation Commission Chapter 210 - Use of Reclaimed Water
http://www.tnrcc.state.tx.us/oprd/rules/pdflib/210_ind.pdf

Utah Administrative Code R317-4
http://www.rules.state.ut.gov/publicat/code/r317-00.1htm#T4

# APPENDIX B
# INDUSTRIAL, UTILITY, AND STATE REGULATORY AGENCY SURVEYS

# Non-Potable Water Use Survey

# Industrial/Ecological Users

## Contact Information

| *Please return survey to:* | ***Your Organization*** *(Please correct or change if needed.)* |
|---|---|
| Utility Name | Name: |
| Utility Contact | Industry: SIC Code or |
| Address — ***To be filled in by utility prior to*** | NAICS Code: |
| Phone | Contact Person: |
| Fax | Address: |
| e-mail | Phone: |
| ***Please return by February 1, 2001.*** | e-mail: |

*Non-potable water is defined as water that does not meet Federal and State standards for human consumption. Non-potable water may have constituents that would be harmful to humans (such as microbiological organisms), but might be suitable for other uses. Sources for non-potable water include treated wastewater plant effluent, untreated groundwater, untreated surface water, or partially treated surface water.*

## Industry Information

*Please provide details on the different aspects of your industry.*

1. Manufacturing Process (What is manufactured? How is water used in the process?)

2. Industrial Activity (Please describe activity, such as trucking company, and how water is used.)

3. Commercial Activity (Please describe activity, such as ?? and how water is used.)

4. Ecological/Environmental (Please describe activity, such as wetland creation, restoration, and enhancement or stream-flow augmentation and how water is used.)

Other (Please describe any activity not discussed above that uses water for non-domestic purposes. Do not include locker rooms, cafeterias, restrooms, etc.)

Is there an industry association that could provide information on water use and water quality? Do you have a contact phone number?

# Non-Potable Water Use Survey

# Industrial/Ecological Users

## Water Quality Information

*Please check the general water quality you need (not what you may actually use) for each of your uses.*

| Customer Classification | Less Than Drinking Water | Drinking Water | Provide Additional Treatment Above Drinking Water |
|---|---|---|---|
| Manufacturing | | | |
| Industrial | | | |
| Commercial | | | |
| Ecological/Environmental | | | |
| Other | | | |

If you checked ìLess than Drinking Waterî for any use, please explain.

If you checked ìAdditional Treatmentî for any use, please describe treatment and what is being removed and added.

*Please list any specific water quality requirements* needed for your use.*

| Customer Classification | Disinfection | TSS | TDS | Nutrients | Sulfates | Color | Hardness | Chlorine | Other |
|---|---|---|---|---|---|---|---|---|---|
| Manufacturing | | | | | | | | | |
| Industrial | | | | | | | | | |
| Commercial | | | | | | | | | |
| Ecological/ Environmental | | | | | | | | | |
| Other | | | | | | | | | |

* TSS = Total Suspended Solids TDS = Total Dissolved Solids Nutrients = Nitrogen, Ammonia, or Phosphorous

# Non-Potable Water Use Survey

# Industrial/Ecological Users

## Water Use Information

*Please circle the closest range of your current total water user.*

| < 5,000 gallons per day | 5,000 to 50,000 gallons per day | 50,000 – 250,000 gallons per day | 250,000 – 500,000 gallons per day | > 1,000,000 gallons per day |
|---|---|---|---|---|

Is any water provided from a source other than the (***Utility Name Here***)? ☐ Yes ☐ No

If Yes, is this other source non-potable? ☐ Yes ☐ No

*How is your water use distributed? Please circle the closest percentage range.*

| | | | | | |
|---|---|---|---|---|---|
| Manufacturing | 0-20 | 20-40 | 40-60 | 60-80 | 80-100 |
| Industrial | 0-20 | 20-40 | 40-60 | 60-80 | 80-100 |
| Commercial | 0-20 | 20-40 | 40-60 | 60-80 | 80-100 |
| Ecological/Environmental | 0-20 | 20-40 | 40-60 | 60-80 | 80-100 |
| Other | 0-20 | 20-40 | 40-60 | 60-80 | 80-100 |

*What time of year do you use the water? Please circle all that apply.*

| | | | | | |
|---|---|---|---|---|---|
| Manufacturing | Dec. – Feb. | Mar. – May | June – Aug. | Sept. – Nov. | All |
| Industrial | Dec. – Feb. | Mar. – May | June – Aug. | Sept. – Nov. | All |
| Commercial | Dec. – Feb. | Mar. – May | June – Aug. | Sept. – Nov. | All |
| Ecological/Environmental | Dec. – Feb. | Mar. – May | June – Aug. | Sept. – Nov. | All |
| Other | Dec. – Feb. | Mar. – May | June – Aug. | Sept. – Nov. | All |

*Please circle where you use or could use non-potable water?*

| | | |
|---|---|---|
| Manufacturing | Currently Use Non-Potable | Could Use Non-Potable |
| Industrial | Currently Use Non-Potable | Could Use Non-Potable |
| Commercial | Currently Use Non-Potable | Could Use Non-Potable |
| Ecological/Environmental | Currently Use Non-Potable | Could Use Non-Potable |
| Other | Currently Use Non-Potable | Could Use Non-Potable |

Thank you for your time. We appreciate the information you have provided. The information will be used to help us and utilities across the country to make more efficient use of our water resources.

May we contact you for additional information or clarification? ☐ Yes ☐ No

Would you like a phone call to discuss the benefits of non-potable water use? ☐ Yes ☐ No

# Non-Potable Water Use Survey

## State Regulatory Agencies

### Contact Information

| *Please return survey to:* | **State Information** (*Please correct or change if needed.*) |
|---|---|
| Mary Stahl<br>Boyle Engineering<br>165 South Union Blvd., #200<br>Lakewood, CO 80228<br>mwstahl@boyleengineering.com<br>**Please return by February 1, 2001.** | Agency Name:<br>Contact Person:<br>Address:<br>Phone:<br>e-mail:<br>***To be filled in prior to mailing.*** |

### Water Use Information

Estimated average day water production for state. ____________ mgd

Estimated demand served by permitted non-potable utilities: Irrigation ____________ % or mgd

Cooling Water____________ % or mgd

Mainline Commercial uses ____________ % or mgd

### Types of Non-Potable Users

*Please list the types of customers served by non-potable water in your state. If you have a database or other form of the information, you may attach it or send it (electronic format preferred).*

| Customer Classification | Types of Customers | Controlling Water Quality Parameters | Contact Information |
|---|---|---|---|
| Manufacturing | | | |
| Industrial | | | |

# Non-Potable Water Use Survey
# State Regulatory Agencies

**Type of Non-Potable Customers (continued)**

| Customer Classification | Types of Customers | Controlling Water Quality Parameters | Contact Information |
|---|---|---|---|
| Commercial | | | |
| Ecological/ Environmental | | | |
| Other *(Please specify)* | | | |

**Non-Potable User Information**

***Please provide information of the numbers and demands of users. If you have a database or other form of the information, you may attach it or send it (electronic format preferred).***

| Customer Classification | No. of Customers | Time of Year | Time of Day | Average Day Use (mgd) | Peak Day Use (mgd) |
|---|---|---|---|---|---|
| Manufacturing | | | | | |
| Industrial | | | | | |
| Commercial | | | | | |
| Ecological/ Environmental | | | | | |
| Other | | | | | |

# Non-Potable Water Use Survey

# State Regulatory Agencies

## Non-Potable Water Quality Information

*Please provide information on the water quality required by non-potable water users in your state. If you have a database or other form of the information, you may attach it or send it (electronic format preferred). Please attach a summary of regulatory requirements for water quality for different users as stated in your regulations.*

| Customer Classification | Disinfection | TSS | TDS | Nutrient Removal | Other Specific Requirements (*Please List*) |
|---|---|---|---|---|---|
| Manufacturing | | | | | |
| Industrial | | | | | |
| Commercial | | | | | |
| Ecological/ Environmental | | | | | |
| Other | | | | | |

**Comments:**

______________________________________________

______________________________________________

______________________________________________

## Other

Do you have information on water quality for non-potable users not currently permitted in your state?
☐ Yes ☐ No If Yes, please attach to returned survey.

Do you have regulations on non-potable use from sources other than treated wastewater? ☐ Yes ☐ No
If Yes, please attach to returned survey.

Please attach your stateís regulation on non-potable reuse with treated wastewater as the source.

Are there additional people we should contact in your state government? ☐ Yes ☐ No
If Yes, please provide name and phone number.

## Additional Comments or Information

# Non-Potable Water Use Survey
# Water Utilities

## Contact Information

| *Please return survey to:* | **Utility Information** (*Please correct or change if needed.)* |
|---|---|
| Mary Stahl | Utility Name: |
| Boyle Engineering | Contact Person: ***To be filled in prior to mailing.*** |
| 165 South Union Blvd., #200 | Address: |
| Lakewood, CO 80228 | Phone: |
| mwstahl@boyleengineering.com | e-mail: |
| **Please return by February 1, 2001.** | |

## Water Use Information

Average day potable water production _________ mgd.

Do you currently have a non-potable program? ☐ Yes ☐ No

If Yes, what is demand served? ___________ mgd

If No, do you have plans for a non-potable program? ☐ Yes ☐ No

Does another utility (i.e., the wastewater treatment provider) in your service area have a non-potable program? ☐ Yes ☐ No

If Yes, what is the demand served? ___________ mgd

## Types of Non-Potable Customers

*Please list the types of customers you serve or plan to serve and the controlling water quality parameters. Please list contact information where acceptable for us to call for more information.*

| Customer Classification | Types of Customers | Controlling Water Quality Parameters | Contact Information |
|---|---|---|---|
| Manufacturing | | | |
| Industrial | | | |

# Non-Potable Water Use Survey
# Water Utilities

## Type of Non-Potable Customers

| Customer Classification | Types of Customers | Controlling Water Quality Parameters | Contact Information |
|---|---|---|---|
| Commercial | | | |
| Ecological/ Environmental | | | |
| Other *(Please specify)* | | | |

## Non-Potable Customer Information

*Please provide information on the numbers and demand of your customers. If in the planning stages, please provide planning numbers.*

| Customer Classification | No. of Customers | Time of Year | Time of Day | Average Day Use (mgd) | Peak Day Use (mgd) |
|---|---|---|---|---|---|
| Manufacturing | | | | | |
| Industrial | | | | | |
| Commercial | | | | | |
| Ecological/ Environmental | | | | | |
| Other | | | | | |

# Non-Potable Water Use Survey

## Water Utilities

### Non-Potable Water Quality Information

*Please provide information on the quality of non-potable water your customers specifically requested.*

| Customer Classification | Disinfection | TSS | TDS | Nutrient Removal | Other Specific Requirements (*Please List*) |
|---|---|---|---|---|---|
| Manufacturing | | | | | |
| Industrial | | | | | |
| Commercial | | | | | |
| Ecological/ Environmental | | | | | |
| Other | | | | | |

**Comments:**

________________________________________

________________________________________

________________________________________

### Other

Do you have information on water quality for customers not served by non-potable? ☐ Yes ☐ No

If Yes, please attach to returned survey.

### Additional Comments or Information

# APPENDIX C
# COMPILED ACCESS DATABASES

All of the survey database queries were done in Access 2000, using a cross-tab query count, which results in a chart showing the number of times a particular variable occurred for each category listed in the first column. Most of these queries were done based on a Customer Classification, or Industry Classification, search: the number of times a variable was listed in each category of Industry, Manufacturing, Commercial, Ecological/Environmental, and Other Irrigation, was then listed in the following columns.

## Utility and State Regulatory Agencies Type and Quantification of Use Survey Results

| Type of Agency | Class_Id | Name | City | Production | State | Types of Customers | Number of Customers | Controlling Water Quality Parameters | Time of Year | Other Requirements | Nutrient Removal | TDS | TSS |
|---|---|---|---|---|---|---|---|---|---|---|---|---|---|
| Utility | 1 | City of San Diego | San Diego | 220 | CA | Electric components | 1 | Title 22 CA Administrative Code and DOHSR | All Year | | Not a concern | <1000 | Not a concern |
| Utility | 2 | City of San Diego | San Diego | 220 | CA | Sand and gravel mining and processing | 1 | Title 22 CA Administrative Code and DOHSR | All Year | | Not a concern | <1000 | Not a concern |
| Utility | 2 | City of San Diego | San Diego | 220 | CA | Commercial building, offices, health care, school | | Title 22 CA Administrative Code and DOHSR | | | Not a concern | <1000 | Not a concern |
| Utility | 5 | City of San Diego | San Diego | 220 | CA | Golf courses, landscaped freeways, parkland, commercial and industrial park, landscaping | 186 | Title 22 CA Administrative Code and DOHSR | All Year | | Not a concern | <1000 | Not a concern |
| Utility | 4 | Las Virgenes Municipal Water District | Calabasas | 17 | CA | Stream flow augmentation for steelhead trout | | Tertiary water meeting Title 22 Contact STDS | | | ~3 PO4, 13 NO3 | | ~ 750 mg/l |
| Utility | 5 | Las Virgenes Municipal Water District | Calabasas | 17 | CA | Outdoor irrigation residential to commercial | ~500 metered connections | Teritary water meeting Title 22 contact STDS | All Year | | ~3 PO4, 13 NO3 | | ~ 750 mg/l |
| Utility | 2 | City of Apopka | Apopka | 6428 | FL | Construction, metal shops | 3 | | | | | Average of 389.000 mg/l | Not more than 20 mg/l |
| Utility | 3 | City of Apopka | Apopka | 6428 | FL | School, nurseries, golf courses | 9 | | | | | Average of 389.000 mg/l | Not more than 20 mg/l |
| Utility | 5 | City of Apopka | Apopka | 6428 | FL | Holding ponds, medians, residential, subdivisions | 856 | | | | | Average of 389.000 mg/l | Not more than 20 mg/l |
| Utility | 2 | Reedy Creek Improvement District | Bay Lake | 20 | FL | Cooling tower makeup, vehicle washing | | PO4, NO3, pH, TDS, Pathogens | All Year | Cooling tower makeup desires constant phosphorus level in feed water, and desires zero ammonia and nitrates - < 2.0 mg/l | | | |
| Utility | 3 | Reedy Creek Improvement District | Bay Lake | 20 | FL | Golf courses, landscape irrigation | | TSS, Pathogens | Mar-May | Cooling tower makeup desires constant phosphorus level in feed water, and desires zero ammonia and nitrates - < 2.0 mg/l | | | |

* Cust Id
1 = Manufacturing
2 = Industrial Activity
3 = Commercial Activity
4 = Ecological/Environmental
5 = Other

(continued)

## Utility and State Regulatory Agencies Type and Quantification of Use Survey Results

| Type of Agency | Class_Id | Name | City | Production | State | Types of Customers | Number of Customers | Controlling Water Quality Parameters | Time of Year | Other Requirements | Nutrient Removal | TDS | TSS |
|---|---|---|---|---|---|---|---|---|---|---|---|---|---|
| Utility | 4 | Reedy Creek Improvement District | Bay Lake | 20 | FL | Wetland hydration | | This is planned, not currently practiced | | Cooling tower makeup desires constant phosphorus level in feed water, and desires zero ammonia and nitrates - < 2.0 mg/l | | | |
| Utility | 5 | Reedy Creek Improvement District | Bay Lake | 20 | FL | Fire protection | | TSS | | Cooling tower makeup desires constant phosphorus level in feed water, and desires zero ammonia and nitrates - < 2.0 mg/l | | | |
| Utility | | City of San Buenaventura | Ventura | 17 | CA | | | | | | | | |
| Utility | 4 | T. P. Smith | Tallahassee | 33 | FL | Developing reuse program for golf course irrigation - 0.20 mgd | Unknown | High level disinfection, TSS removal via filtration to meet TSS-5 ppm limit, total N-<10 ppm | Jun-Aug | These are general requirements - not in use yet | Total -< 10 ppm | 3000 ppm | 5 ppm |
| Utility | 5 | T. P. Smith | Tallahassee | 33 | FL | Agricultural reuse 17.074 mgd | Two sites - 2,163 ac and 85 ac | CBOD, TSS-20 ppm, groundwater nitrate - 10 ppm | All Year | These are general requirements - not in use yet | Total -< 10 ppm | 3000 ppm | 5 ppm |
| Utility | | City of Tallahassee | Tallahassee | 35 | FL | | | | | | | | |
| Utility | 2 | Denver Water Department | Denver | 206 | CO | Xcel, Conoco | | Phosphorus, ammonia | | The information provided is in addition to what you have for our reuse project. You should have the zoo information already. | The potential customers asked for low levels but did not specify what the levels should be | | |
| Utility | 4 | Denver Water Department | Denver | 206 | CO | Washington Park, City Park, Denver Zoo | 3 | Phosphorus, ammonia, chlorine | All Year | The information provided is in addition to what you have for our reuse project. You should have the zoo information already. | The potential customers asked for low levels but did not specify what the levels should be | | |
| Utility | 1 | West Basin Municipal Water District | | 133 | | Petroleum industry, Chevron, BP ASARCO, Exxon Mobile | 3 | Cooling - low to no ammonia, boilers - very pure, essentially distilled H20 | Sep-Nov | Low ammonia, demineralized | X | X | X |
| Utility | 4 | West Basin Municipal Water District | | 133 | | Microfiltered reverse osmosis water (MFRO) to the West Coast Barrier | 1 | | All Year | Low ammonia, demineralized | X | X | X |

* Cust Id
1 = Manufacturing
2 = Industrial Activity
3 = Commercial Activity
4 = Ecological/Environmental
5 = Other

(continued)

## Utility and State Regulatory Agencies Type and Quantification of Use Survey Results

| Type of Agency | Class_Id | Name | City | Production | State | Types of Customers | Number of Customers | Controlling Water Quality Parameters | Time of Year | Other Requirements | Nutrient Removal | TDS | TSS |
|---|---|---|---|---|---|---|---|---|---|---|---|---|---|
| Utility | 5 | West Basin Municipal Water District | | 133 | | Irrigation - cemeteries, landscaping, residential, schools | | | Jun-Aug | Low ammonia, de-mineralized | X | X | X |
| Utility | 1 | Irvine Ranch Water District | Irvine | 0 | CA | Dye plant | 1 | Consistent water quality more than wq | | Consistency | | | |
| Utility | 2 | Irvine Ranch Water District | Irvine | 0 | CA | Construction - dust control | 1 | | | Consistency | | | |
| Utility | 3 | Irvine Ranch Water District | Irvine | 0 | CA | Toilets - urban high rises | 5 | Color - aesthetic quality | | Consistency | | | |
| Utility | 5 | Irvine Ranch Water District | Irvine | 14.3 | CA | Irrigation - urban landscaping and agriculture | 2,200 | | Jun-Aug | Consistency | | | |
| Utility | 2 | South Bermuda WWTP (FL) | | 0 | | 2 power plants - cooling towers, Florida Power, KVA Kissimmee | 2 | | Jun-Aug | See attached | | | <0.5 |
| Utility | 5 | South Bermuda WWTP (FL) | | 0 | | 2 golf courses, irrigation | 2 | | Jun-Aug | See attached | | | <0.5 |
| Utility | 5 | South Bermuda WWTP (FL) | | 0 | | Irrigation | 9 golf courses, medians, 2 schools, 2 churches, 3-4 parks | | Mar-May | See attached | | | <0.5 |
| Utility | 3 | Naples Waste Treatment Plant | | 0 | | 200 wash cages | 1200 wash | | All Year | | | | |
| State | 3 | Arizona Department of Environmental Quality | Phoenix | 82 | AZ | Golf course, agricultural, landscape, toilets | 77 | Fecal coliform | | See Reclaimed Water Quality Standards | | | |
| State | 5 | Arizona Department of Environmental Quality | Phoenix | 82 | AZ | Recreational, residential irrigation, schools, parks, residential - toilets | 63 | | | See Reclaimed Water Quality Standards | | | |
| State | 2 | NDEP | Carson City | | NV | Industrial parks, cooling towers (2) power plants, gravel plant and construction, dust & soil conditioning | Several | CFU/MPN/100 ml, 2.2 fecal coliform, 2nd treatment at a minimum or 2.2 total coliform with tertiary treatment of effluent | Jun-Aug | Minimum of 2ndary treatment | Yes Some | | |
| State | 3 | NDEP | Carson City | | NV | Golf courses, agricultural, dust & soil conditioning, and surface/drip irrigation with controlled access | about 50-60 statewide | CFU/MPN 2.2 fecal coliform 2ndary treatment minimum, see regulations, 22/100 ml fecal coliform minimum | Mar-May | Minimum of 2ndary treatment | Yes Some | | |
| State | 4 | NDEP | Carson City | | NV | Water impoundments with body contact, wildlife management areas, wetlands - treating "grey water", high BOD for irrigation | 3/4 | 2.2 total coliform 2ndary treatment, 2.2 fecal coliform, site specified | All Year | Minimum of 2ndary treatment | Yes Some | | |

* Cust Id
1 = Manufacturing
2 = Industrial Activity
3 = Commercial Activity
4 = Ecological/Environmental
5 = Other

(continued)

## Utility and State Regulatory Agencies Type and Quantification of Use Survey Results

| Type of Agency | Class_Id | Name | City | Production | State | Types of Customers | Number of Customers | Controlling Water Quality Parameters | Time of Year | Other Requirements | Nutrient Removal | TDS | TSS |
|---|---|---|---|---|---|---|---|---|---|---|---|---|---|
| State | 5 | NDEP | Carson City | | NV | Parkway/median irrigation, fire academy for exercises, sports complexes, cemetary, treated wastewater disposal by irrigation-spray | 1 | 2.2 fecal coliform, 2.2 total coliform, 200/100 with buffer zones or 2.2 with no buffer | All Year | Minimum of 2ndary treatment | Yes Some | | |
| State | 2 | NDEP | Carson City | | NV | Industrial parks, cooling towers (2) power plants, gravel plant and construction, dust & soil conditioning | Several | CFU/MPN/100 ml, 2.2 fecal coliform, 2nd treatment at a minimum or 2.2 total coliform with tertiary treatment of effluent | Jun-Aug | Minimum of 2ndary treatment | Yes | | |
| State | 3 | NDEP | Carson City | | NV | Golf courses, agricultural, dust & soil conditioning, and surface/drip irrigation with controlled access | about 50-60 statewide | CFU/MPN 2.2 fecal coliform 2ndary treatment minimum, see regulations, 22/100 ml fecal coliform minimum | Mar-May | Minimum of 2ndary treatment | Yes | | |
| State | 4 | NDEP | Carson City | | NV | Water impoundments with body contact, wildlife management areas, wetlands - treating "grey water", high BOD for irrigation | 3/4 | 2.2 total coliform 2ndary treatment, 2.2 fecal coliform, site specified | All Year | Minimum of 2ndary treatment | Yes | | |
| State | 5 | NDEP | Carson City | | NV | Parkway/median irrigation, fire academy for exercises, sports complexes, cemetary, treated wastewater disposal by irrigation-spray | 1 | 2.2 fecal coliform, 2.2 total coliform, 200/100 with buffer zones or 2.2 with no buffer | All Year | Minimum of 2ndary treatment | Yes | | |
| State | 2 | Division of Water Quality Rules | Salt Lake City | 2290 | UT | Cooling water | | | | Case-by-case | | | |
| State | 5 | Division of Water Quality Rules | Salt Lake City | 2290 | UT | Agricultural irrigation | | BOD, TSS, coliform | | Case-by-case | | | |
| State | | Florida Department of Environmental Protection | Tallahassee | 18180 | FL | | | | | | | | |
| State | 2 | Colorado Department of Public Health and Environment | Denver | 0 | CO | Rawhide power plant, cooling tower | 1 | | All Year | Ammonia reduction, phosphous | | | |

* Cust Id
1 = Manufacturing
2 = Industrial Activity
3 = Commercial Activity
4 = Ecological/Environmental
5 = Other

(continued)

# Utility and State Regulatory Agencies Type and Quantification of Use Survey Results

| Type of Agency | Class_Id | Name | City | Production | State | Types of Customers | Number of Customers | Controlling Water Quality Parameters | Time of Year | Other Requirements | Nutrient Removal | TDS | TSS |
|---|---|---|---|---|---|---|---|---|---|---|---|---|---|
| State | 4 | Colorado Department of Public Health and Environment | Denver | 0 | CO | Denver SE Suburban Water and Sanitary, and Cherry Creek Basin | 1 | | All Year | Ammonia reduction, phosphous | | | |
| State | 5 | Colorado Department of Public Health and Environment | Denver | 0 | CO | Landscape irrigation | Several | *E. coli* and TSS or Turbidity | Jun-Aug | Ammonia reduction, phosphous | | | |
| State | 1 | Maryland Department of Environment | | 0 | | Mining (coal), pump water, reuse their own, dust control, slurry disposal, ??? Washing, settling ponds 90% own recycle | | | | | | | |
| State | 2 | Maryland Department of Environment | | 0 | | Power plant - 2 mgd cooling | | | | | | | |
| State | 3 | Maryland Department of Environment | | 0 | | Car washes - more than 50% reuse their own water | | | | | | | |
| State | 4 | Maryland Department of Environment | | 0 | | Ponds, pilot program, Frederick Department of P.W. County wetlands ww treatment | | | | | | | |
| State | 5 | Maryland Department of Environment | | 0 | | Irrigation, 3 golf courses | | | | | | | |
| State | 2 | Oregon Department of Environmental Quality | | 0 | | 4836 mgd/year of reuse H20 going to l and application - food processing, wash water | | | | | | | |
| State | 4 | Oregon Department of Environmental Quality | | 0 | | Wetlands | | | | | | | |
| State | 4 | Oregon Department of Environmental Quality | | 0 | | Irrigation - 77 wwt facilities for land application, 2,727 mg/year of reuse from domestic treated sources | | | | | | | |

* Cust Id
1 = Manufacturing
2 = Industrial Activity
3 = Commercial Activity
4 = Ecological/Environmental
5 = Other

## Industrial/Ecological Users Water Quality Information Survey Answers

| Cust ID | Ind/Eco Name | City | ST | SIC Code | WQ Needed? | Explanation | Disinfection | TSS | TDS | Nutrients | Sulfates | Color | Hardness |
|---|---|---|---|---|---|---|---|---|---|---|---|---|---|
| 5 | Olivas Adobe Vista Parks Department | Ventura | CA | | Both Less than Drinking Water and Drinking Water | Used for landscape irrigation | | | | | | | |
| 1 | A. L. Johnson Company | | | 3365 | Drinking Water | Water used to make molding material must be potable water | | | | | | | |
| 1 | Allied Fasteners, Inc. | Camarillo | CA | | Drinking Water | Domestic | | | | | | | |
| 5 | SCI | | | | Drinking Water | | | | | | | | |
| 1 | Siemens Solar Industries L.P. | Canaullo | | 3674 | < Drinking Water | I believe we could use less than drinking water grade for most of our manufacturing needs, but I'm not sure on our RO plant | | X | X | | X | | X |
| 4 | Tierra Rejada Golf Club | Moorpark | CA | | < Drinking Water | This water would be used to water turf | | | X | X | | | |
| 1 | El Paso Electric Company | El Paso | TX | | Above Drinking Water | 4 cooling towers use tertiary treated waste-water only. Boiler makeup water quality requires R.O. D.J. high purity treatment | | | 1200 mmhos | | | | 180 ppm |
| 5 | The Church of Jesus Christ of Latter Day Saints | | CA | | < Drinking Water | Used for landscaping | | | | | | | |
| 4 | City of Calabasas | | | | Both Less than Drinking Water and Drinking Water | Reclaimed water has some salinity issues related to soil structure (growth medium) and drainage but so does drinking water. The availability and cost of drinking water prohibits any further exploration of "better than" for irrigation (landscape purposes | | X | X | | | | X |
| 5 | County Line Shopping Center | Thousand Oaks | CA | | Both Less than Drinking Water and Drinking Water | Irrigation | | | | | | | |
| 3 | Creekside Calabasas HOA | Camarillo | CA | | | | | | | | | | |

* Cust Id
1 = Manufacturing
2 = Industrial Activity
3 = Commercial Activity
4 = Ecological/Environmental
5 = Other

(continued)

# Industrial/Ecological Users Water Quality Information Survey Answers

| Cust ID | Ind/Eco Name | City | ST | SIC Code | WQ Needed? | Explanation | Disinfection | TSS | TDS | Nutrients | Sulfates | Color | Hardness |
|---|---|---|---|---|---|---|---|---|---|---|---|---|---|
| 4 | Greenbrier Properties, Inc. | Westlake Village | CA | | Both Less than Drinking Water and Drinking Water | Pond and foundations | | | | | | | |
| 4 | Helmet House | Calabasas Hills | CA | | Both Less than Drinking Water and Drinking Water | Irrigation and for employees | | | | | | | |
| 5 | La Ventura Properties | Thousand Oaks | CA | | Both Less than Drinking Water and Drinking Water | Irrigation | | | | | | | |
| 5 | Lakeshore Oak View Ranch, Regency Hills | W. V. | CA | | Both Less than Drinking Water and Drinking Water | Common area landscape irrigation and 3 swimming pools | | | | | | | |
| 3 | The Kanan Group | Agoura Hills | CA | | < Drinking Water | Irrigation | | | | | | | |
| 4 | Calabasas Golf and Country Club | Calabasas | | | Both Less than Drinking Water and Drinking Water | Irrigation of hybrid bermuda grass tees, rough and fairways | | | <768 | OK | <200 | | |
| 5 | Soka University of America | | | | < Drinking Water | Landscape irrigation | | | | | | | |
| 3 | South Hills Property Owners Association | Thousand Oaks | CA | | | | | | | | | | |
| 3 | Stonecrest Patio Homeowners Association | Camarillo | CA | | < Drinking Water | Used for landscape areas only | | | | | | | |
| 3 | Watergate Property Owners | | | | | | | | | | | | |
| 5 | City of Westlake Village | | | | Both Less than Drinking Water and Drinking Water | For landscape irrigation | | | | | | | |
| 3 | Westlake Village Inn | | | | Both Less than Drinking Water and Drinking Water | Less than drinking water is used for irrigation of landscape | | | | | | | |
| 1 | Dollar Rent A Car | Denver | CO | | < Drinking Water | F.A.C. | | | | | | | |
| 2 | General Chemical | Denver | CO | 2819 | Drinking Water | F.A.C. | | | | | | | |
| 5 | Hertz Corporation | Denver | CO | | < Drinking Water | F.A.C. | | | | | | | |
| 2 | Samsonite Corporation | Denver | CO | | < Drinking Water | Treated with bioxides, fungicides, solids dispersant, anti-foaming agents, corrosion, unit 4 condenser, sulfuric acid, pH of 7.0 | No biologicals | <10,000 ppm | | | | | |

* Cust Id
1 = Manufacturing
2 = Industrial Activity
3 = Commercial Activity
4 = Ecological/Environmental
5 = Other

(continued)

# Industrial/Ecological Users Water Quality Information Survey Answers

| Cust ID | Ind/Eco Name | City | ST | SIC Code | WQ Needed? | Explanation | Disinfection | TSS | TDS | Nutrients | Sulfates | Color | Hardness |
|---|---|---|---|---|---|---|---|---|---|---|---|---|---|
| 2 | Weyerhaeuser | Denver | CO | 2653 | < Drinking Water | Biocides, moving to ozone control | X | X | X | | | | |
| 1 | Xcel Energy-Cherokee Plant | | | | Both Less than Drinking Water and Drinking Water | Cooling water is recycled | | | | | | | |
| 2 | J. G. Management Company, Inc. | | | | < Drinking Water | | | X | | | | | |
| 5 | Monroe County Detention Center | | | | < Drinking Water | | | X | | | | | |
| 1 | Vero Beach Power Plant | | | | Drinking Water | See attached for limits | Low organics (see attached), no amenes, low metals | | | | | | Total: <450 mg/l (a Hardness <375 mg/l |
| 2 | Electric Power Research Institute | St. Louis | | | Drinking Water | | | | | | | | |
| 4 | Total Petroleum | Commerce City | CO | | Drinking Water | | | | | | | | |
| 5 | Florida Department of Corrrections | | | | Drinking Water | | | | | | | | |
| 1 | Sand Creek Chemical | Denver | CO | | | | X | | | Nutirent-free | | | 30 ppm CaC03 |
| 4 | Rocky Mountain Arsenal | Denver | CO | | Drinking Water | | | | | | | | |
| 5 | General Mail Facility | | | | Drinking Water | | | | | | | | |
| 4 | Park Hill Golf Course/Clayton College | Denver | CO | | < Drinking Water | | | | | | | | |
| 5 | Gateway Park/Sand Creek Metro District | | | | < Drinking Water | | | | | | | | |
| 1 | Denver International Airport | Denver | CO | | < Drinking Water | | Low BOD, no fecal coli | | | | | | ALK <300 mg/l |
| 2 | Denver Bulk Mail Center | Denver | CO | | Both Less than Drinking Water and Drinking Water | | No fecal coliforms | | | | | | |
| 5 | Brannan S&G | | | | < Drinking Water | | Title 22 | | | | | | |
| 2 | ASARCO | Leadville | CO | | | Only needed for cooling/sanitary use, sanitary is pretreated, and cooling water is treated with lungicide/biocides and rust/corrison inhibitors | | | | | | | |

* Cust Id
1 = Manufacturing
2 = Industrial Activity
3 = Commercial Activity
4 = Ecological/Environmental
5 = Other

(continued)

## Industrial/Ecological Users Water Quality Information Survey Answers

| Cust ID | Ind/Eco Name | City | ST | SIC Code | WQ Needed? | Explanation | Disinfection | TSS | TDS | Nutrients | Sulfates | Color | Hardness |
|---|---|---|---|---|---|---|---|---|---|---|---|---|---|
| 5 | Conoco | | | | | Only needed for cooling/ sanitary use, sanitary is pre-treated, and cooling water is treated with lungicide/biocides and rust/corrison inhibitors | | | 100-1200 umhos | | | | |
| 1 | Tallahassee's Purdem Power Plant | | | | | | | | | | | | |
| 1 | Florida Power, Inc. | | | | Both Less than Drinking Water and Drinking Water | The boilers need DW - quality water, but the cooling towers could use non-potable | X | | | | | | |
| 2 | Public Service Company - Cherokee Plant | Denver | CO | | < Drinking Water | Boiler water - 99.99% pure; de-mineralized and F.A.C. and DEP bleach | X | | | | | | X |
| 2 | Lakeland McIntosh Power Plant | | | | < Drinking Water | The injection process requires: filtration, reverse osmosis, demineralization | X | | Careful | | | | |
| 1 | Royalty Carpets | | | | Drinking Water | See attached | <25 mg/l (ppm) | <25 mg/l, 3-5 mg/l | 500-700 mg/l | | | | 150-250 ppm as CaCo3 Ca CaCO3 = 130-200 ppm |
| 2 | Dillon Dairy | Denver | CO | | < Drinking Water | | X | 5 ppm | | | | | |
| 5 | | | | | < Drinking Water | | X | 5 ppm | | | | | |
| 2 | | | | | < Drinking Water | Filtration | | | | | | | |
| 1 | | | | | < Drinking Water | | | | | | | | Softened for manufacturing |
| 5 | | | | | < Drinking Water | | | | | | | | |
| 1 | | | | | Above Drinking Water | Some water is chlorinated above drinking water specs. We add chlorine to sanitize equipment. | | | | | | | |

* Cust Id
1 = Manufacturing
2 = Industrial Activity
3 = Commercial Activity
4 = Ecological/Environmental
5 = Other

(continued)

# Industrial/Ecological Users Water Quality Information Survey Answers

| Ind/Eco Name | Chlorine | Other | Industry Information | Water Use | Other Source | Other SourceNP | Distribution | Time of Year | Non_Pot_Use |
|---|---|---|---|---|---|---|---|---|---|
| Olivas Adobe Vista Parks Department | | | Landscape irrigation | < 5,000 gpd | FALSE | FALSE | 80 - 100 | All Year | Currently Use Non-Potable |
| A. L. Johnson Company | | | Manufacturer of aluminum of castings. Water is mixed with molding material to create molds in which to pour molten metal. | < 5,000 gpd | FALSE | FALSE | 80 - 100 | All Year | |
| Allied Fasteners, Inc. | | | | | FALSE | FALSE | | | |
| SCI | | | We don't manufacture. We sell wholesale kitchenware. We use water for drinking, cleaning, toilets. We have a sprinkler system that automatically waters our shrubbery. | < 5,000 gpd | FALSE | FALSE | | All Year | |
| Siemens Solar Industries L.P. | X | | Solar cells, wafering cleaning, etching, RO water making cooling towers, scrubbers. Landscape. Semiconductor association (possible). | 50,000 - 250,000 gpd | FALSE | FALSE | 80 - 100 | All Year | Could Use Non-Potable |
| Tierra Rejada Golf Club | | | | 250,000 - 500,000 gpd | FALSE | FALSE | 80 - 100 | All Year | Could Use Non-Potable |
| El Paso Electric Company | .7 ppm (free) | 1.2 ppm ortho phosphoric 7.85 ph | Production of electricity. Water used for cooling purposes to condense steam and return condensate to the boiler. Water also used for boiler make-up, cooling of bearings. | >1,000,000 gpd | TRUE | TRUE | 80 - 100 | All Year | Currently Use Non-Potable |
| The Church of Jesus Christ of Latter Day Saints | | | Church property used for landscape irrigation. | < 5,000 gpd | FALSE | FALSE | 80 - 100 | All Year | Currently Use Non-Potable |
| City of Calabasas | X | | Used as an irrigation water source for installing and maintaing plant materials. Unknown. However I have thoughts on salinity, nutrients and other plant sensitivity issues regarding the use of reclaimed water. | >1,000,000 gpd | FALSE | FALSE | 80 - 100 | All Year | Currently Use Non-Potable |
| County Line Shopping Center | | | Irrigation of plants. | < 5,000 gpd | FALSE | FALSE | 80 - 100 | All Year | Currently Use Non-Potable |
| Creekside Calabasas HOA | | | Irrigation only for HOA. | | FALSE | FALSE | | | |
| Greenbrier Properties, Inc. | | | Irrigation and man-made pond. | < 5,000 gpd | FALSE | FALSE | 20 - 40 | All Year | |
| Helmet House | | | Irrigation. | | FALSE | FALSE | 80 - 100 | All Year | Currently Use Non-Potable |
| La Ventura Properties | | | Irrigation. | < 5,000 gpd | FALSE | FALSE | 80 - 100 | All Year | Currently Use Non-Potable |
| Lakeshore Oak View Ranch, Regency Hills | | | Common area irrigation only. | | FALSE | FALSE | | All Year | Currently Use Non-Potable |
| The Kanan Group | | | Retail shopping center. Water for bathrooms and restaurant uses. Irrigation is gray water. | | FALSE | FALSE | | All Year | Currently Use Non-Potable |

* Cust Id
1 = Manufacturing
2 = Industrial Activity
3 = Commercial Activity
4 = Ecological/Environmental
5 = Other

(continued)

# Industrial/Ecological Users Water Quality Information Survey Answers

| Ind/Eco Name | Chlorine | Other | Industry Information | Water Use | Other Source | Other SourceNP | Distribution | Time of Year | Non_Pot_Use |
|---|---|---|---|---|---|---|---|---|---|
| Calabasas Golf and Country Club | Chloride <100 | | Irrigation of golf course. Pace Consulting 618-272-9897. | 250,000 - 500,000 gpd | FALSE | FALSE | 80 - 100 | All Year | Currently Use Non-Potable |
| Soka University of America | | Reclaimed water | Landscaping and irrigation. | 5,000 - 50,000 gpd | FALSE | FALSE | 80 - 100 | All Year | Currently Use Non-Potable |
| South Hills Property Owners Association | | | Landscape irrigation. No records are kept. | | FALSE | FALSE | | | |
| Stonecrest Patio Homeowners Association | | | HOA - landscape only. | | FALSE | FALSE | 80 - 100 | All Year | Currently Use Non-Potable |
| Watergate Property Owners | | | Irrigation only for HOA. | | FALSE | FALSE | | | |
| City of Westlake Village | | | Landscape irrigation. Much of landscape irrigation water drains onto streets, drains that discharge into a lake that overflows to a stream. Watershed Advisory Grp. 650-853-0454; Environmental Water Management, Fax 818-707-1509; Biosolutions 818-991-9997 | | FALSE | FALSE | 80 - 100 | All Year | Currently Use Non-Potable |
| Westlake Village Inn | | | Water is used for irrigation of landscaping for 17 acres. The property has a manmade lake, water is used for a stream and is recycled. Las Virgenes Municipal Water Dist. 818-251-2100, Scott Harris | 5,000 - 50,000 gpd | TRUE | FALSE | 40 - 60 | All Year | Could Use Non-Potable |
| Dollar Rent A Car | | | Car rental company; washing cars we use a reclaim car wash system. | < 5,000 gpd | FALSE | FALSE | 60 - 80 | Jun - Aug | |
| General Chemical | | | We manufacture liquid aluminum sulfate. Water is the main raw material used in our process. Our product is used to filter water and wastewaters. It must meet AWWA specs. | 5,000 - 50,000 gpd | FALSE | FALSE | 80 - 100 | All Year | |
| Hertz Corporation | | | Washing cars. We do have a reclaim system in which we reuse the water in the car wash. | 5,000 - 50,000 gpd | FALSE | FALSE | 80 - 100 | All Year | Currently Use Non-Potable |
| Samsonite Corporation | | | Parks washer-discharge to ???-metal finisher. Manufacturing facility-metal finisher. Cooling towers for boiler plant and industrial cooling. Colorado Water Commission | < 5,000 gpd | FALSE | FALSE | 60 - 80 | All Year | Could Use Non-Potable |
| Weyerhaeuser | | | Corrugated containers. Steam boiler for paper machine. Manufacturing. Lawn irrigation. | 5,000 - 50,000 gpd | FALSE | FALSE | 40 - 60 | All Year | Could Use Non-Potable |

* Cust Id
1 = Manufacturing
2 = Industrial Activity
3 = Commercial Activity
4 = Ecological/Environmental
5 = Other

(continued)

# Industrial/Ecological Users Water Quality Information Survey Answers

| Ind/Eco Name | Chlorine | Other | Industry Information | Water Use | Other Source | Other SourceNP | Distribution | Time of Year | Non_Pot_Use |
|---|---|---|---|---|---|---|---|---|---|
| Xcel Energy-Cherokee Plant | 1 | Conductivity <5500 umhos (cooling), <3500 umhos (boiler), diss. Fe <0.1 ppm, Na must be low | Electricity production. Water used for heat rejection cooling and for steam generation. Cooling towers-heat removal via heat exchangers. Boiler water-steam generation. Water discharged to river after use (primarily) for cooling. | >1,000,000 gpd | TRUE | TRUE | 80 - 100 | All Year | Currently Use Non-Potable |
| J. G. Management Company, Inc. | | | | 5,000 - 50,000 gpd | TRUE | FALSE | 20 - 40 | All Year | Currently Use Non-Potable |
| Monroe County Detention Center | | | Chilling, toilets, fires, limited irrigation - permitted but not used | | FALSE | FALSE | | | |
| Vero Beach Power Plant | <100 mg/l, <49 mg/l | | Cooling tower, blowdown to WWTP, tiny bit of irrigation (3%) | 500,000 - 1,000,000 gpd | TRUE | FALSE | 40 - 60 | All Year | Currently Use Non-Potable |
| Electric Power Research Institute | | 7 mg/l nitrate | Cooling tower | 500,000 - 1,000,000 gpd | TRUE | FALSE | 40 - 60 | | Could Use Non-Potable |
| Total Petroleum | | 7 mg/l nitrate, 0.5 mg/l phosphorus | Cooling - recycled already - 173 kgal/day, boiler feed - 86 kgal/day, process -desalting crude oil uses - 49 kgal/day | 5,000 - 50,000 gpd | FALSE | FALSE | 80 - 100 | Jun - Aug | Currently Use Non-Potable |
| Florida Department of Corrrections | | 7 mg/l nitrate | Flushing toilets, washing clothes, irrigation | 5,000 - 50,000 gpd | FALSE | FALSE | 0 - 20 | Jun - Aug | Currently Use Non-Potable |
| Sand Creek Chemical | | | Steam generation, process make-up, cooling | 50,000 - 250,000 gpd | FALSE | FALSE | 80 - 100 | Sep - Nov | Currently Use Non-Potable |
| Rocky Mountain Arsenal | | | Dust control, revegetation/remediation/ wetlands, irrigation | 50,000 - 250,000 gpd | FALSE | FALSE | 80 - 100 | All Year | Could Use Non-Potable |
| General Mail Facility | | | Cooling units/swamp coolers, Tital 22 - irrigation | 250,000 - 500,000 gpd | FALSE | FALSE | 20 - 40 | All Year | Currently Use Non-Potable |
| Park Hill Golf Course/Clayton College | | Very low phosphorus necessary | Well augmentation, irrigation | 250,000 - 500,000 gpd | FALSE | FALSE | 20 - 40 | All Year | Currently Use Non-Potable |
| Gateway Park/Sand Creek Metro District | | | Lakes - 20 acres with 15 proposed, irrigation | 250,000 - 500,000 gpd | FALSE | FALSE | 80 - 100 | All Year | Could Use Non-Potable |
| Denver International Airport | | PO4 <0.2, conducicivity <150 umho | Plant heating and cooling, car wahses - 18,000 gal/day - buses, di-icing facilities 900,000 fal/year - must be potable, rentals. Irrigation and wetlands - reuse potential | >1,000,000 gpd | FALSE | FALSE | 0 - 20 | All Year | Could Use Non-Potable |
| Denver Bulk Mail Center | | | Sanitary use 300,000 outflow/month - 10,000 gal/day, irrigation and cooling (summer only) 1,400,000 gallons/ month - 46,600 gal/day | >1,000,000 gpd | FALSE | FALSE | 40 - 60 | All Year | |

| | |
|---|---|
| * Cust Id | 3 = Commercial Activity |
| 1 = Manufacturing | 4 = Ecological/Environmental |
| 2 = Industrial Activity | 5 = Other |

(continued)

## Industrial/Ecological Users Water Quality Information Survey Answers

| Ind/Eco Name | Chlorine | Other | Industry Information | Water Use | Other Source | Other SourceNP | Distribution | Time of Year | Non_Pot_Use |
|---|---|---|---|---|---|---|---|---|---|
| Brannan S&G | | <0.5 mg/l PO4, <7 mg/l NO3-N | Settlement ponds, lime plant - 1,000 gal/day (peak 1,200 gal/day), dust control - roads, Stack scrubbers | >1,000,000 gpd | FALSE | FALSE | 40 - 60 | All Year | Could Use Non-Potable |
| ASARCO | | | Cooling towers, wash cars/trucks, wet down piles of waste/product, irrigation of local lawns | 50,000 - 250,000 gpd | FALSE | FALSE | 60 - 80 | All Year | |
| Conoco | | | Cooling water - 864,000 gal/day maximum (576-720 kgal/day average) , boiler feed water - 187,000 gal/day - potable - w/hot line for silica removal, followed by sodium cation exchange, reclamation/irrigation, | 50,000 - 250,000 gpd | FALSE | FALSE | 0 - 20 | Jun - Aug | Currently Use Non-Potable |
| Tallahassee's Purdem Power Plant | | | Cooling towers, boiler water - evaporated H20 | >1,000,000 gpd | FALSE | FALSE | 20 - 40 | All Year | |
| Florida Power, Inc. | | 0.1 ppm ammonia | Intercession City - control nitrogen oxide emission - injected into combustion zone, Hines Energy complex - Cooling pond and make-up water | >1,000,000 gpd | FALSE | FALSE | 60 - 80 | All Year | Could Use Non-Potable |
| Public Service Company - Cherokee Plant | | | Electricity, boiler, cooling towers, pump seals and washdown | 50,000 - 250,000 gpd | FALSE | FALSE | 20 - 40 | All Year | Could Use Non-Potable |
| Lakeland McIntosh Power Plant | | All very low | 2 cooling towers | 50,000 - 250,000 gpd | FALSE | FALSE | 60 - 80 | All Year | Currently Use Non-Potable |
| Royalty Carpets | | < 2mg/l PO4 or <2 ppm | Carpet - reuse water - after dyeing. Lawn watering. Dyeing - Softeners. Industrial grade treated. Carpet and Rug Institute (Dalton, GA) Mundy Hall 626-967-5268. | >1,000,000 gpd | FALSE | FALSE | 20 - 40 | All Year | Could Use Non-Potable |
| Dillon Dairy | 1.0 ppm | 2.0 ppm | Dairy food processing plant. FDA and USDA will not allow non-potable water. | >1,000,000 gpd | FALSE | FALSE | 20 - 40 | Jun - Aug | Could Use Non-Potable |

* Cust Id
1 = Manufacturing
2 = Industrial Activity
3 = Commercial Activity
4 = Ecological/Environmental
5 = Other

(continued)

## Industrial/Ecological Users Water Quality Information Survey Answers

| Ind/Eco Name | Chlorine | Other | Industry Information | Water Use | Other Source | Other SourceNP | Distribution | Time of Year | Non_Pot_Use |
|---|---|---|---|---|---|---|---|---|---|
| | 1.0 ppm | 2.0 ppm | | >1,000,000 gpd | FALSE | FALSE | 40 - 60 | Jun - Aug | Currently Use Non-Potable |
| | | low phosphorus | | 5,000 - 50,000 gpd | FALSE | FALSE | 80 - 100 | All Year | Could Use Non-Potable |
| | | pH - 6.5 -7.5 | | 5,000 - 50,000 gpd | FALSE | FALSE | 20 - 40 | All Year | Could Use Non-Potable |
| | | | | 5,000 - 50,000 gpd | FALSE | FALSE | 60 - 80 | All Year | Currently Use Non-Potable |
| | | | | 5,000 - 50,000 gpd | FALSE | FALSE | 0 - 20 | | Could Use Non-Potable |
| | | | | 5,000 - 50,000 gpd | FALSE | FALSE | | | |
| | | | | >1,000,000 gpd | FALSE | FALSE | 80 - 100 | All Year | Could Use Non-Potable |
| | | | | >1,000,000 gpd | FALSE | FALSE | 0 - 20 | Jun - Aug | |
| | | | | 50,000 - 250,000 gpd | TRUE | TRUE | 80 - 100 | All Year | Currently Use Non-Potable |
| | | | | >1,000,000 gpd | TRUE | TRUE | 80 - 100 | Jun - Aug | Currently Use Non-Potable |
| | | | | >1,000,000 gpd | FALSE | FALSE | 80 - 100 | All Year | Could Use Non-Potable |
| | | | | 250,000 - 500,000 gpd | TRUE | FALSE | 20 - 40 | All Year | Currently Use Non-Potable |
| | | | | 250,000 - 500,000 gpd | FALSE | FALSE | 20 - 40 | All Year | Currently Use Non-Potable |
| | | | | >1,000,000 gpd | FALSE | FALSE | 80 - 100 | All Year | Currently Use Non-Potable |
| | | | | 500,000 - 1,000,000 gpd | FALSE | FALSE | 0 - 20 | All Year | Currently Use Non-Potable |
| | | | | 250,000 - 500,000 gpd | FALSE | FALSE | 60 - 80 | All Year | Currently Use Non-Potable |
| | | | | 5,000 - 50,000 gpd | FALSE | FALSE | 80 - 100 | All Year | Could Use Non-Potable |

* Cust Id
1 = Manufacturing
2 = Industrial Activity
3 = Commercial Activity
4 = Ecological/Environmental
5 = Other

# APPENDIX D
# SUMMARY TABLES

**Table D.1**
**Suggested guidelines for water reuse[1]**

| Types of Reuse | Reclaimed Water Quality[2] | Reclaimed Water Monitoring | Setback Distances[3] | Comments |
|---|---|---|---|---|
| ***Urban Reuse*** | pH = 6 - 9 | pH - weekly | 50 ft (15 m) to potable water supply wells | At controlled-access irrigation sites where design and operational measures significantly reduce the potential of public contact with reclaimed water, a lower level of treatment, e.g., secondary treatment and disinfection to achieve ≤14 fecal coli/100 ml, may be appropriate. |
| All types of landscape irrigation, (e.g., golf courses, parks, cemeteries) also vehicle washing, toilet flushing, use in fire protection systems and commerical air conditioners, and other uses with similar access or exposure to the water | 5-10 mg/L BOD | BOD - weekly | | Chemical (coagulant and/or polymer) addition prior to filtration may be necessary to meet water quality recommendations. |
| | ≤ 2 NTU | Turbidity - continuous | | The reclaimed water should not contain measurable levels of pathogens. |
| | No detectable fecal coli/100 m[7,8] | Coliform - daily | | Reclaimed water should be clear, odorless, and contain no substances that are toxic upon ingestion. |
| | 1 mg/L $Cl^2$ residual (min.) | $Cl_2$ residual - continuous | | A higher chlorine residual and/or a longer contact time may be necessary to assure that viruses and parasites are inactivated or destroyed. |
| | | | | A chlorine residual of 0.5 mg/l or greater in the distribution system is recommended to reduce odors, slime, and bacterial re-growth. |

(continued)

**Table D.1 (Continued)**

| Types of Reuse | Reclaimed Water Quality[2] | Reclaimed Water Monitoring | Setback Distances[3] | Comments |
|---|---|---|---|---|
| ***Restricted Access Area Irrigation*** | pH = 6 - 9 | pH - weekly | 300 ft (90 m) to potable water supply wells | |
| | ≤ 30 mg/L BOD[5] | BOD - weekly | | |
| Sod farms, silviculture sites, and other areas where public access is prohibited, restricted, or infrequent | 30 mg/L SS | SS - daily | 100 ft (30 m) to areas accessible to the public (if spray irrigation) | If spray irrigation, SS less than 30 mg/L may be necessary to avoid clogging of sprinkler heads. |
| | 200 fecal[7, 3] coli/100 mL | Coliform - daily | | |
| | 1 mg/L $Cl_2$ residual (min.) | $Cl_2$ residual - continuous | | |
| ***Recreational Impoundments*** | pH 6-9 | pH - weekly | 500 ft (150 m) to potable water supply wells (minimum) if bottom not sealed | Dechlorination may be necessary to protect aquatic species of flora and fauna. |
| | 10 mg/L BOD[5] | BOD - weekly | | Reclaimed water shall be non-irritating to skin and eyes. |
| | ≤ 2 NTU | Turbidity - continuous | | Nutrient removal may be necessary to avoid algae growth in impoundments. |
| | No detectable fecal coli/100 ml[7,8] | Coliform - daily | | Reclaimed water should be clear, odorless, and contain no substances that are toxic upon ingestion. |
| | 1 mg/L $C1_2$ residual (min.)[9] | $Cl_2$ residual - continuous | | Chemical (coagulant and/or polymer) addition prior to filtration may be necessary to may meet water quality recommendations. |
| | | | | The reclaimed water should not contain measurable levels of pathogens. |
| | | | | A higher chlorine residual and/or a longer contact time may be necessary to assure that viruses and parasites are inactivated or destroyed. |
| | | | | Fish caught in impoundments can be consumed. |

(continued)

**Table D.1 (Continued)**

| Types of Reuse | Reclaimed Water Quality[2] | Reclaimed Water Monitoring | Setback Distances[3] | Comments |
|---|---|---|---|---|
| ***Landscape Impoundments*** | 30 mg/L BOD[5] | pH - weekly | 500 ft (150 m) to potable water supply wells (minimum) if bottom not sealed | Nutrient removal processes may be necessary to avoid algae growth in impoundments. |
| Aesthetic impoundment where public contact with reclaimed water is not allowed | < 30 mg/L SS | SS - daily | | Dechlorination may be necessary to protect aquatic species of flora and fauna. |
| | ≤ 200 fecal coli/ 100 ml[7,11,12] | Coliform - daily | | |
| | 1 mg/L $Cl_2$ residual (min.)[9] | $Cl_2$ residual - continuous | | |
| ***Construction Uses*** | < 30 mg/L BOD | BOD - weekly | | Worker contact with reclaimed water should be minimized. |
| Soil compaction, dust control, washing aggregate, making concrete | < 30 mg/L SS | SS - daily | | A higher level of disinfection, e.g., to achieve 5-14 fecal coli/100 mL, should be provided where frequent worker contact with reclaimed water is likely. |
| | 5-200 fecal coli/ 100 mL [7,11,12] | Coliform - daily | | |
| | 1 mg/L $Cl_2$ residual (min.)[9] | $Cl_2$ residual - continuous | | |
| ***Industrial Reuse*** | pH = 6 - 9[5] | pH - daily | 300 ft (90 m) to areas accessible to the public | Windblown spray should not reach areas accessible to users or the public. |
| Once-through cooling, recirculating cooling towers | 5-30 mg/L BOD | BOD - weekly | | |
| | 5-30 mg/L SS | SS - weekly | 300 ft (90 m) to areas accessible to the public | Windblown spray should not reach areas accessible to the public. |
| | 5-200 fecal coli/ 100 ml[7,11,12] | Coliform - daily<br>$Cl_2$ residual - continuous | May be reduced if high level of disinfection is provided | Additional treatment by user is usually provided to prevent scaling, corrosion, biological growths, fouling and foaming. |
| | $Cl_2$ mg/L residual (min.)[9] | | | |
| | Variable, depends on recirculation ratio | | | |

(continued)

**Table D.1 (Continued)**

| Types of Reuse | Reclaimed Water Quality[2] | Reclaimed Water Monitoring | Setback Distances[3] | Comments |
|---|---|---|---|---|
| Other Industrial Uses | | | | |
| ***Environmental Reuse***<br>Wetlands, marshes, wildlife habitat, stream augmentation | Variable, but not to exceed:<br>≤ 30 mg/L BOY<br>< 30 mg/L SS<br>≤ 200 fecal[7,11,12] coli/100 ml | Depends on site specific use<br>BOD - weekly<br>SS - daily<br>Coliform - daily<br>$Cl_2$ residual - continuous | | Dechlorination may be necessary to protect aquatic species of flora and fauna.<br>Possible effects on groundwater should be evaluated.<br>Receiving water quality requirements may necessitate additional treatment.<br>The temperature of the reclaimed water should not adversely affect ecosystem. |
| Groundwater recharge | Site specific and use dependent | Depends on treatment and use | Site specific | Facility should be designed to ensure that no reclaimed water reaches potable water supply aquifers. |
| By spreading or injection into nonpotable aquifers | | | | For injection projects, filtration and disinfection may be needed to prevent clogging. |
| ***Indirect Potable Reuse***<br>Groundwater recharge by spreading into potable aquifers<br>Groundwater recharge by injection into potable aquifers | Site specific<br>Meet drinking water standards after percolation through vadose zone | Includes, but not limited to, the following:<br>PH - daily<br>Coliform - daily | 2,000 ft (600 m) to extraction wells. May vary depending on treatment provided and site-specific conditions. | The depth to groundwater (i.e., thickness of the vadose zone) should be at least 6 feet (2 m) at the maximum groundwater mounding point.<br>The reclaimed water should be retained underground for at least 1 year prior to withdrawal. |

(continued)

**Table D.1 (Continued)**

| Types of Reuse | Reclaimed Water Quality[2] | Reclaimed Water Monitoring | Setback Distances[3] | Comments |
|---|---|---|---|---|
| ***Indirect Potable Reuse*** (continued) | Includes, but not limited to, the following:<br>pH = 6.5 - 8.5<br>≤ 2 NTU[4]<br>No detectable fecal coli/100 ml[7,8]<br>1 mg/L $Cl_2$ residual (min.)<br>Meet drinking water standards | $Cl_2$ residual - continuous<br>Drinking water standards - quarterly[5] | 2000 ft (600 m) to extraction wells. May vary depending on site-specific conditions | Recommended treatment is site-specific and depends on factors such as type of soil, percolation rate, thickness of vadose zone, native groundwater quality, and dilution.<br>Monitoring wells are necessary to detect the influence of the recharge operation on the groundwater. |
| | | Other depends on constituent.<br>Includes, but not limited to, the following:<br>pH - daily<br>Turbidity continuous<br>Coliform - daily<br>$Cl_2$ residual - continuous<br>Drinking water standards - quarterly | | The reclaimed water should not contain measurable levels of pathogens after percolation through the vadose zone.[10]<br>The reclaimed water should be retained underground for at least 1 year prior to withdrawal.<br>Monitoring wells are necessary to detect the influence of the recharge operation on the groundwater.<br>Recommended quality limits should be met at the point of injection. |

(continued)

**Table D.1 (Continued)**

| Types of Reuse | Reclaimed Water Quality[2] | Reclaimed Water Monitoring | Setback Distances[3] | Comments |
|---|---|---|---|---|
| ***Indirect Potable Reuse*** (continued) | | Other[5] depends on constituent | | The reclaimed water should not contain measurable levels of pathogens at the point of injection.[10] |
| | | | | A higher chlorine residual and/or a longer contact time may be necessary to assure virus inactivation. |
| ***Indirect Potable Water*** | Includes, but not limited to, the following: | Includes, but not limited to, the following: | Site specific | Recommended level of treatment is site-specific and depends on factors such as receiving water quality, time and distance to point of withdrawal, dilution and subsequent treatment prior to distribution for potable uses. |
| Augmentation of surface supplies | pH = 6.5 – 8.5 | pH - daily | | |
| | ≤ 2 NTU[6] | Turbidity - continuous | | |
| | No detectable fecal coli/100 ml[7,8] | Coliform - daily | | The reclaimed water should not contain measurable levels of pathogens.[10] |
| | 1 mg/L $Cl_2$ residual (min.)[9] | $Cl_2$ residual - continuous | | A higher chlorine residual and/or a longer contact time may be necessary to assure virus inactivation. |
| | Meet drinking water standards | Drinking water standards - quarterly | | |
| | | Other - depends on constituent | | |

*Source:* EPA Guidelines for Water Reuse, 1992

1. These guidelines are based on water reclamation and reuse practices in the U.S., and they are especially directed at states that have not developed their own regulations or guidelines. While the guidelines should be useful in many areas outside the U.S., local conditions may limit the applicability of the guidelines in some countries. It is explicitly stated that the direct application of these suggested guidelines will not be used by AID as strict criteria for funding.
2. Unless otherwise noted, recommended quality limits apply to the reclaimed water at the point of discharge from the treatment facility.
3. Setback distances are recommended to protect potable water supply sources from contamination and to protect humans from unreasonable health risks due to exposure to reclaimed water.
4. Disinfection means the destruction, inactivation, or removal of pathogenic microorganisms by chemical, physical, or biological means. Disinfection may be accomplished by chlorination, ozonation, other chemical disinfectants, UV radiation, membrane processes, or other processes.
5. As determined from the 5-day BOD test.
6. The recommended turbidity limit should be met prior to disinfection, The average turbidity should be based on a 24-hour time period. The turbidity should not exceed 5 NTU at any time. If SS is used in lieu of turbidity, the average SS should not exceed 5 mg/l.
7. Unless otherwise noted, recommended coliform limits are median values determined from the bacteriological results of the last 7 days for which analyses have been completed. Either the membrane filter or fermentation tube technique may be used.
8. The number of fecal coliform organisms should not exceed 14/100 mL in any sample.
9. Total chlorine residual after a minimum contact time of 30 minutes.
10. It is advisable to fully characterize the microbiological quality of the reclaimed water prior to implementation of a reuse program.
11. The number of fecal coliform organisms should not exceed 800/100 mL in any sample.
12. Some stabilization pond systems may be able to meet this coliform limit without disinfection.

**Table D.2**
**Access and allowed use across different states, as defined in regulations in Spring 2001**
**(Please check with state for updates and specific regulatory language.)**

| State, access level | Definition of access, allowed uses, and setbacks |
|---|---|
| Arizona (unrestricted) | Irrigation of open areas, including playgrounds, lawns, parks, golf courses, and other areas where public access is not restricted. |
| Arizona (restricted) | Irrigation of orchard crops and crops not subject to rotation, which minimizes contact of the reclaimed wastewater with the fruit or foliage. Industrial wastewater is not to be used for this purpose. |
| | Irrigation of pastures where incidental ponding or standing water is prevented. |
| | Irrigation of landscaped areas, cemeteries, highway medians, golf courses, and other areas where public access is restricted. Golf courses in areas that are separated by a fence or barrier of at least four feet. |
| | Industrial use to be considered on an individual basis, but use in cooling systems is allowed. |
| California (undisinfected secondary recycled water) | Surface irrigation of the following: (1) Orchards where the recycled water does not come into contact with the edible portion of the crop, (2) Vineyards where the recycled water does not come into contact with the edible portion of the crop, (3) Non food-bearing trees (Christmas tree farms are included in this category provided no irrigation with recycled water occurs for a period of 14 days prior to harvesting or allowing access by the general public), (4) Fodder and fiber crops and pasture for animals not producing milk for human consumption, (5) Seed crops not eaten by humans, (6) Food crops that must undergo commercial pathogen-destroying processing before being consumed by humans, and (7) Ornamental nursery stock and sod farms provided no irrigation with recycled water occurs for a period of 14 days prior to harvesting, retail sale, or allowing access by the general public. |
| | Recycled water used for flushing sanitary sewers. |
| | 150 ft. irrigation setback from domestic water supplies and wells. |
| California (disinfected secondary-2.2 recycled water) | Surface irrigation of food crops where the edible portion is produced above ground and not contacted by the recycled water. |
| | Supply for restricted recreational impoundments and for any publicly accessible impoundments at fish hatcheries. |
| | 100 ft. irrigation setback from domestic water supplies and wells. |

(continued)

**Table D.2 (Continued)**

| State, access level | Definition of access, allowed uses, and setbacks |
|---|---|
| California (disinfected secondary-23 recycled water) | Surface irrigation of the following: (1) Cemeteries, (2) Freeway landscaping, (3) Restricted access golf courses, (4) Ornamental nursery stock and sod farms where access by the general public is not restricted, (5) Pasture for animals producing milk for human consumption, and (6) Any nonedible vegetation where access is controlled so that the irrigated area cannot be used as if it were part of a park, playground or school yard. |
| | Landscape impoundments that do not utilize decorative fountains. |
| | Recycled water for industrial or commercial cooling or air conditioning that does not involve the use of a cooling tower, evaporative condenser, spraying, or any mechanism that creates a mist. |
| | Recycled water may be used in:(1) Industrial boiler feed, (2) Nonstructural fire fighting, (3) Backfill consolidation around non-potable piping, (4) Soil compaction, (5) Mixing concrete, (6) Dust control on roads and streets, (7) Cleaning roads, sidewalks and outdoor work areas and (8) Industrial process water that will not come into contact with workers. |
| | 100 ft. irrigation setback from domestic water supplies and wells. |
| California (disinfected tertiary recycled water) | Surface irrigation of: (1) Food crops, including all edible root crops, where the recycled water comes into contact with the edible portion of the crop, (2) Parks and playgrounds, (3) School yards, (4) Residential landscaping, (5) Unrestricted access golf courses, and (6) Any other irrigation use not specified in this section and not prohibited by other sections of the California Code of Regulations. |
| | Additionally, disinfected tertiary recycled water used for irrigation, or soil that has been irrigated with recycled water, is allowed to come into contact with the edible portion of food crops eaten raw by humans. |
| | Non-restricted recreational impoundments (subject to conventional treatment). |
| | Non-restricted recreational impoundments without conventional treatment, provided the recycled water is monitored for the presence of pathogenic organisms in accordance with section 60305.1 (*Giardi*a, enteric viruses, and *Cryptosporidium* screening standards). |

(continued)

**Table D.2 (Continued)**

| State, access level | Definition of access, allowed uses, and setbacks |
|---|---|
| | Recycled water used for industrial or commercial cooling or air conditioning that involves the use of a cooling tower, evaporative condenser, spraying or any mechanism that creates a mist. Provided a drift eliminator shall be used whenever the cooling system is in operation, and a chlorine, or other, biocide shall be used to treat the cooling system recirculating water to minimize the growth of *Legionella* and other micro-organisms.) |
| | Recycled water can be used for the following: (1) Flushing toilets and urinals, (2) Priming drain traps, (3) Industrial process water that may come into contact with workers, (4) Structural fire fighting, (5) Decorative fountains, (6) Commercial laundries, (7) Consolidation of backfill around potable water pipelines, (8) Artificial snow making for commercial outdoor use, and (9) Commercial car washes, including hand washes if the recycled water is not heated, where the general public is excluded from the washing process. |
| | 50 ft. irrigation setback from domestic water supplies with conditions; 100 ft. setback without certain conditions. |
| Colorado (all-inclusive) | Reclaimed domestic wastewater shall be confined to the authorized use area. Runoff of reclaimed domestic wastewater shall be strictly minimized. Direct and windblown spray shall be confined to the area designated and approved for application. Precautions shall be taken to ensure that reclaimed domestic wastewater will not be sprayed on any facility or area not designated for application such as occupied buildings, domestic drinking water facilities, or facilities where food is being prepared for human consumption. Notification shall be provided to inform the public that reclaimed domestic wastewater is being used for irrigation and is not safe for drinking. The notification shall include posting of signs of sufficient size to be clearly read in all application areas and around impoundments with appropriate wording (e.g., ATTENTION: RECLAIMED DOMESTIC WASTEWATER - DO NOT DRINK) in the dominant language(s) expected to be spoken at the site. |
| | There shall be no impoundment or irrigation of reclaimed domestic wastewater within 100 feet of any well used for domestic supply unless, in the case of an impoundment, the impoundment is lined with a synthetic material with a permeability of $10^{-6}$ cm/sec or less. |
| Colorado (unrestricted) | Reclaimed domestic wastewater used for unrestricted landscape irrigation shall at a minimum receive treatment such that it is oxidized, filtered and disinfected. |

(continued)

**Table D.2 (Continued)**

| State, access level | Definition of access, allowed uses, and setbacks |
|---|---|
| Colorado (restricted) | Reclaimed domestic wastewater used for restricted landscape irrigation shall at a minimum receive secondary treatment with disinfection.<br>Irrigation shall occur only during periods approved in the Notice of Authorization so as to strictly minimize public contact with reclaimed domestic wastewater or, where the applicator has installed barriers (as approved in the Notice of Authorization, to prevent public access to a site), irrigation may occur at any time that the barriers are fully secured except that irrigation shall cease one hour prior to the barriers being totally or partially removed.<br>A 500-foot setback distance shall be provided from the edge of the wetted area to potable water supply wells, Class I surface waters; or Class II surface waters.<br>Overland flow land application systems, rapid-rate land application systems providing continuous loading to a single percolation cell, other land application systems involving less than secondary treatment prior to application, septic tanks, and ground water disposal systems using Class I wells injecting effluent or wastes into Class G-IV waters shall be excluded from the definition of reuse.<br>For a period of 15 days from the last application of reclaimed water, land application areas shall not be used for the grazing of cattle whose milk is intended for human consumption. No waiting period is required if high-level disinfection. |
| Florida (secondary treatment – surface water disposal, groundwater injection) | If subsurface application systems are used, no setback distances to buildings are required. If subsurface application systems are used, the setback distance to the site property line shall be reduced to 30 feet. |
| Florida (secondary treatment, new plants) | Reclaimed water may be used for once-through cooling. |

(continued)

**Table D.2 (Continued)**

| State, access level | Definition of access, allowed uses, and setbacks |
|---|---|
| Florida (secondary treatment & basic disinfection) | Overland flow systems and certain underdrained slow-rate land application systems as specified. |
| | Slow-rate land application systems shall maintain a distance of 100 feet from the edge of the wetted area to buildings that are not part of the treatment facility, utilities system, or municipal operation; or to the site property line. This distance shall be reduced to 50 feet if the setback is vegetated with trees or shrubs to create a continuous visual barrier at least five feet high to minimize aerosol drift. |
| | This distance shall be reduced to 50 feet if only low trajectory, low pressure nozzles or surface application techniques are used within the outermost 50 feet of the application area. |
| | Reclaimed water may be used for open cooling towers, if a 300' setback is maintained from the property line. |
| | Reclaimed water may be used for toilet flushing in commercial or industrial facilities or buildings. Reclaimed water may be used for toilet flushing in motels, hotels, apartment buildings, and condominiums where the individual guests or residents do not have access to the plumbing system for repairs or modifications. Reclaimed water shall not be used for toilet flushing in any residential property or dwelling unit where the residents have access to the plumbing system for repairs or modifications. |
| | Reclaimed water may be used to provide water for fire protection. Reclaimed water may be supplied to fire hydrants. Hydrants supplied with reclaimed water shall be color-coded and shall have no connection to the potable water supply. |
| | Reclaimed water may be used to provide water for fire protection in sprinkler systems located in commercial or industrial facilities or buildings. Reclaimed water may be used to provide water for fire protection in sprinkler systems located in motels, hotels, apartment buildings, and condominiums where the individual guests or residents do not have access to the plumbing system for repairs or modifications. Such sprinkler systems shall be color-coded and shall be supplied only by reclaimed water. |
| | Reclaimed water may be used for dust control at construction sites. |
| | Reclaimed water may be used for indoor or outdoor aesthetic purposes. Examples of aesthetic uses of reclaimed water include decorative pools, fountains, ponds, lagoons, and other aesthetic features. |
| | Additional uses that are approved include: (a) Water supply for commercial laundries. (b) Vehicle washing. (c) Flushing of sanitary sewers and reclaimed water lines. (d) Mixing of concrete. (e) Manufacture of ice for ice rinks. (f) Cleaning roads, sidewalks, and outdoor work areas. |

(continued)

**Table D.2 (Continued)**

| State, access level | Definition of access, allowed uses, and setbacks |
|---|---|
| Florida (secondary treatment & intermediate disinfection) | Differing setbacks are allowed for different purposes. Generally, minimum is 75 feet form the edge of the wetland area to potable water supply wells. |
| Florida (secondary treatment & high-level disinfection) | Wetlands may be used for the discharge or treatment of reclaimed water or effluent. Public access to the wetlands shall be restricted. |
| | If subsurface application systems are used and if high-level disinfection is provided, the setback distance to the site property line shall be reduced to 10 feet. |
| | Setback distance shall be 50 feet from the edge of the wetted area to buildings that are not part of the treatment facility, utilities system, or municipal operation; or to the site property line. |
| Hawaii | No current standards for applications of reclaimed wastewater exist. The following guidelines for classifications of reuse water are in place: |
| | R-1 – highest quality recycled water. This water has gone through filtration and disinfection. It is safe for use on lawns, golf courses, parks, and other places that people frequent. |
| | R-2 – slightly lower quality recycled water. R-2 is secondary (biologically) treated wastewater, which has also been disinfected. Its use requires more caution and restrictive controls than R-2 wastewater. |
| | R-3 – the least pure class of recycled water. R-3 quality water is wastewater that has been treated at the secondary level. It can only be used for irrigation at places where people rarely go. |
| | Wherever applicable, the director shall use the effluent requirements for non-domestic wastewater as set forth by the EPA. The director may specify coliform requirements, which are more or less stringent than the requirements of subsection 11-62-26(4). These requirements are to be based on public accessibility, methods of effluent reuse and public health considerations. |
| | Recycled water is used for dust control, agricultural irrigation, and the irrigation of golf courses and other landscapes. |
| Maryland | No reclaimed water standard exists at this time. Standards for land application of wastewater, solid waste disposal, and point discharge of effluent (NPDES permit) are currently required for the different applications of reclaimed water. |

(continued)

**Table D.2 (Continued)**

| State, access level | Definition of access, allowed uses, and setbacks |
|---|---|
| Maryland (DRAFT) | Reclaimed water may be used for irrigation of: farmland; golf courses; athletic fields; turf; landscaping; and any other use that the department considers appropriate.<br>The Department may establish buffer and setback requirements for the use of reclaimed water … as follows: from potable wells and surface water intakes, up to 100 feet; from intermittent and perennial streams and residential structures, up to 25 feet; and from schools and playgrounds, up to 50 feet. |
| Nevada (general) | Any person using treated effluent for irrigation shall post a notice at the site of irrigation warning the general public to avoid contact with the treated effluent.<br>Except as otherwise provided in this subsection, a person shall not use treated effluent to irrigate crops for human consumption. A person may use treated effluent for surface irrigation of fruit bearing trees and nut bearing trees.<br>A person using treated effluent to irrigate by sprinklers shall conduct the irrigation in a manner that inhibits the treated effluent from drifting or carrying outside the buffer zone.<br>A person shall not allow treated effluent used in irrigation to run off the site being irrigated. |
| Nevada (A) – no public access | Irrigation of pasture; or other agricultural purposes except growing crops for human consumption, where public access to the site being irrigated is prohibited.<br>800 ft. setback from domestic water supplies. |
| Nevada (A)(1) – restricted access | Irrigation of pasture; or other agricultural purposes except growing crops for human consumption, where public access to the site being irrigated is prohibited.<br>400 ft. setback from domestic water supplies. |
| Nevada (B) – limited access | Irrigation of: A golf course, cemetery or greenbelt where public access to the site being irrigated is controlled and human contact with the treated effluent does not occur.<br>An impoundment where all activities are prohibited and human contact with the treated effluent does not occur.<br>100 ft. setback from domestic water supplies. |

(continued)

**Table D.2 (Continued)**

| State, access level | Definition of access, allowed uses, and setbacks |
|---|---|
| Nevada I – unrestricted | Irrigation of: A cemetery, highway median, greenbelt, park, playground or residential or commercial lawn where public access to the site being irrigated is controlled and human contact with the treated effluent cannot reasonably be expected. |
| | Impoundments where full body contact with the treated effluent cannot reasonably be expected. |
| | No setback from domestic water supplies. |
| Oregon (IV – unrestricted) | A body of reclaimed water in which no limitations are imposed on body-contact water recreation activities. |
| | Recreation is limited to fishing, boating, and other non-body-contact water recreation activities. |
| | Surface irrigation of orchards and vineyards where edible portion of crop does not contact the ground, and fruit or nuts shall not be harvested off the ground. |
| | Surface or spray irrigation (with no animals in the area) for processed food crops, sod, ornamental nursery stock, Christmas trees, firewood, and commercial timber. |
| | Golf courses without contiguous residences. |
| | Irrigation of cemeteries, highway medians, landscapes without frequent public access. |
| Oregon (III – restricted) (continued) | Use of reclaimed water in evaporative cooling systems shall be approved only if the user can demonstrate that aerosols will not present a hazard to public health. |
| | Construction/industrial use. |
| | 10 ft. setback from surface irrigation. |
| | 100 ft. setback from food preparation or drinking fountains. |
| Oregon (II – restricted) | Same as Oregon III, but samples are taken are taken weekly, instead of tri-weekly, and the setback from irrigation sprays is 70 ft. |
| | May not be used in an impoundment. |

(continued)

**Table D.2 (Continued)**

| State, access level | Definition of access, allowed uses, and setbacks |
|---|---|
| Oregon (I – prevented) | Fodder, fiber, commercial timber and seed Crops not for unrestricted human ingestion; not to be sprayed within 30 days of harvest. |
| | Irrigation of parks, playgrounds, schoolyards, golf courses with contiguous residences. |
| | Construction/industrial use. |
| | Reclaimed water shall not spray directly on areas of food preparation or drinking fountains. |
| Texas (general) | Reclaimed water shall not be utilized in a way that degrades ground water quality to a degree adversely affecting its actual or potential uses. |
| | Food crops that may be consumed raw by humans shall not be spray irrigated. Food crops including orchard crops that will be substantially processed prior to human consumption may be spray irrigated. Other types of irrigation that avoid contact of reclaimed water with edible portions of food crops are acceptable. |
| Texas (I – unrestricted) | Residential irrigation; Urban uses, including irrigation of public parks, golf course with unrestricted public access, school yards, or athletic fields; use of reclaimed water for fire protection, either in internal sprinkler systems or external fire hydrants; irrigation of food crops where the applied reclaimed water may have direct contact with the edible part of the crop, unless the food crop undergoes a pasteurization process; irrigation of pastures for milking animals; maintenance of impoundments or natural water bodies where recreational activities, such as wading or fishing, are anticipated even though the water body was not specifically designed for such a use; toilet or urinal flush water; other similar activities where the potential for unintentional human exposure may occur. |

(continued)

**Table D.2 (Continued)**

| State, access level | Definition of access, allowed uses, and setbacks |
|---|---|
| Texas (II – restricted – non-pond) | Irrigation of sod farms, silviculture, limited access highway rights of way, and other areas where human access is restricted or unlikely to occur. The restriction of access to areas under irrigation with reclaimed water could include the following: (I ) The irrigation site is considered to be remote. (ii) The irrigation site is bordered by walls or fences and access to the site is controlled by the owner/operator of the irrigation site. (iii) The irrigation site is not used by the public during the times when irrigation operations are in progress. Such sites may include golf courses, cemeteries, and landscaped areas surrounding commercial or industrial complexes. The "syringing" or "wetting" of greens and tees on golf courses shall be allowable under Type II so long as the "syringing" is done with hand-held hoses as opposed to automatic irrigation equipment. The public need not be excluded from areas where irrigation is not taking place. For example, irrigation of golf course fairways at night would not prohibit the use of club house or other facilities located a sufficient distance from the irrigation. (iv) The irrigation site is restricted from public access by local ordinance or law with specific standards to achieve such a purpose. |
| | Irrigation of food crops where the reclaimed water is not likely to have direct contact with the edible part of the crop, or where the food crop undergoes pasteurization prior to distribution for consumption. |
| | Irrigation of animal feed crops other than pasture for milking animals. |
| | Maintenance of impoundments or natural water bodies where direct human contact is not likely. |
| | Soil compaction or dust control in construction areas where application procedures minimize aerosol drift to public areas. |
| | Cooling tower makeup water. Use for cooling towers, which produce significant aerosols adjacent to public access areas, may have special requirements. |
| | Irrigation or other non-potable uses of reclaimed water at a wastewater treatment facility. |

(continued)

**Table D.2 (Continued)**

| State, access level | Definition of access, allowed uses, and setbacks |
|---|---|
| Texas (II – restricted – pond) | See Texas (II-restricted – non-pond ) conditions. |
| Utah (I – human exposure) | Residential irrigation, including landscape irrigation at individual houses. |
| | Urban uses, which include nonresidential landscape irrigation, golf course irrigation, toilet flushing, fire protection, and other uses with similar potential for human exposure. |
| | Irrigation of food crops where the applied reclaimed water is likely to have direct contact with the edible part. Type I water is required for all spray irrigation of food crops. |
| | Irrigation of pasture for milking animals. |
| | Impoundments of wastewater where direct human contact is likely to occur. |
| | 50 ft. setbacks for irrigation, from potable sources; 500 ft. setbacks for impoundment, away from potable sources. |
| Utah (II – no human exposure) | Irrigation of sod farms, silviculture, limited access highway rights of way, and other areas where human access is restricted or unlikely to occur. |
| | Irrigation of food crops where the applied reclaimed water is not likely to have direct contact with the edible part, whether the food will be processed or not (spray irrigation not allowed). |
| | Irrigation of animal feed crops other than pasture used for milking animals. |
| | Impoundments of wastewater where direct human contact is not allowed or is unlikely to occur. |
| | Cooling water. Use for cooling towers, which produce aerosols in populated areas, may have special restrictions imposed. |
| | Soil compaction or dust control in construction areas. |
| | 300 ft. setbacks for irrigation, from potable sources; 500 ft. setbacks for impoundment, away from potable sources. |

# APPENDIX E
# NATIONAL PRIMARY SURFACE DRINKING WATER STANDARDS

**Table E.1**
**National drinking water standards**

| Contaminant | MCLG[1] (mg/L) | MCL or TT[1] (mg/L) | Potential Health Effects From Ingestion of Water | Sources of Contaminant in Drinking Water |
|---|---|---|---|---|
| ***Microorganisms*** | | | | |
| *Cryptosporidium* | Zero | TT | Gastrointestinal illness | Human and fecal animal waste |
| *Giardia lamblia* | Zero | TT | Gastrointestinal illness | Human and fecal animal waste |
| *Legionella* | Zero | TT | Legionnaire's Disease, a type of pneumonia | Found naturally in water; multiplies in heating systems |
| Total Coliforms (including fecal coliform and *E. Coli*) | Zero | 5.0%[2] | Indicates presence of other potentially harmful bacteria | Coliforms are naturally present in the environment; as well as feces; fecal coliforms and *E. coli* only come from human and animal fecal waste |
| Turbidity | N/A | TT | Indicates water quality and filtration effectiveness | Soil runoff |
| Viruses (enteric) | Zero | TT[3] | Gastrointestinal illness | Human and animal fecal waste |
| ***Disinfection Byproducts*** | | | | |
| Bromate | Zero | 0.010 | Increased risk of cancer | Byproduct of drinking water disinfection |
| Chlorite | 0.8 | 1.0 | Anemia; infants and young children: nervous system effects | Byproduct of drinking water disinfection |
| Haloacetic acids (HAA5) | N/A | 0.060 | Increased risk of cancer | Byproduct of drinking water disinfection |
| Total Trihalomethanes (TTHMs) | N/A | 0.080 | Liver, kidney, or central nervous system problems; increased risk of cancer | Byproduct of drinking water |

(continued)

**Table E.1 (Continued)**

| Contaminant | MCLG[1] (mg/L) | MCL or TT[1] (mg/L) | Potential Health Effects From Ingestion of Water | Sources of Contaminant in Drinking Water |
|---|---|---|---|---|
| ***Inorganic Chemicals*** | | | | |
| Antimony | 0.006 | 0.006 | Increase in blood cholesterol; decrease in blood sugar | Discharge from petroleum refineries; fire retardants; ceramics, electronics; solder |
| Arsenic | 0[3] | 0.010 as of 1/23/06 | Skin damage or problems with circulatory systems, and may have increased risk of getting cancer | Erosion of natural deposits; runoff from orchards, runoff form glass and electronics production wastes |
| Asbestos (fiber >10 micrometers) | 7 million fibers per liter | 7 MFL | Increased risk of developing benign intestinal polyps | Decay of asbestos in water mains; erosion of natural deposits |
| Barium | 2 | 2 | Increase in blood pressure | Discharge of drilling wastes; discharge from metal refineries, erosion of natural deposits with beryllium |
| Beryllium | 0.004 | 0.004 | Intestinal lesions | Discharge from metal refineries and coal-burning factories; discharge from electrical, aerospace, and defense industries |
| Cadmium | 0.005 | 0.005 | Kidney damage | Corrosion of galvanized pipes; erosion of natural deposits; discharge from metal refineries; runoff form waste batteries and paints |
| Chromium (total) | 0.1 | 0.1 | Allergic dermatitis | Discharge from steel and pulp mills; erosion of natural deposits |
| Copper | 1.3 | TT[4] Action Level = 1.3 | Short term exposure: Gastrointestinal distress<br>Long term exposure: Liver or kidney damage | Corrosion of household plumbing systems; erosion of natural deposits |

(continued)

Table E.1 (Continued)

| Contaminant | MCLG[1] (mg/L) | MCL or TT[1] (mg/L) | Potential Health Effects From Ingestion of Water | Sources of Contaminant in Drinking Water |
|---|---|---|---|---|
| Cyanide (as free cyanide) | 0.2 | 0.2 | Nerve damage or thyroid problems | Discharge from steel/metal factories; discharge from plastic and fertilizer factories |
| Fluoride | 4.0 | 4.0 | Skeletal and dental fluoresis | Water additive; erosion of natural deposits; discharge from fertilizer and aluminum factories |
| Lead | Zero | TT[4] Action Level = 0.015 | Kidneys, nervous system damage, delayed or impaired mental development in infants and children | Corrosion of household pluming systems; erosion of natural deposits |
| Mercury (inorganic) | 0.002 | 0.002 | Kidney damage | Erosion of natural deposits; discharge form refineries and factories; runoff form landfills and croplands |
| Nitrate (measured as Nitrogen) | 10 | 10 | Methemoglobulinemia | Runoff form fertilizer use; leaching form septic tanks, sewage; erosion of natural deposits |
| Nitrite (measured as Nitrogen) | 1 | 1 | Methemoglobulinemia | Runoff form fertilizer use; leaching form septic tanks, sewage; erosion of natural deposits |
| Selenium | 0.05 | 0.05 | Hair or fingernail loss; numbness in fingers or toes; circulatory problems | Discharge from petroleum refineries; erosion of natural deposits; discharge from mines |
| Thallium | 0.0005 | 0.002 | Hair loss; changes in blood; kidney, intestine, or liver problems | Leaching from ore-processing sites; discharge form electronics, glass, and drug factories |

(continued)

**Table E.1 (Continued)**

| Contaminant | MCLG[1] (mg/L) | MCL or TT[1] (mg/L) | Potential Health Effects From Ingestion of Water | Sources of Contaminant in Drinking Water |
|---|---|---|---|---|
| ***Radionuclides*** | | | | |
| Alpha particles | Zero | 15 (pCi/L) | Increased risk of cancer | Erosion of natural deposits of certain minerals that are radioactive and may emit a form of radiation known as alpha radiation |
| Beta particles and photon emitters | Zero | 4 mrem per year | Increased risk of cancer | Decay of natural and man made deposits of certain minerals that are radioactive and may emit forms of radiation known as photons and beta radiation |
| Radium 226 and Radium 228 (combined) | Zero | 5pCi/L | Increased risk of cancer | Erosion of natural deposits |
| Uranium | Zero | 30 ug/L as of 12/08/03 | Increased risk of cancer, kidney toxicity | Erosion of natural deposits |

*Source:* USEPA, 2002.

Organics are not listed as they are not addressed by reuse regulations and were not an issue per the literature search or the surveys.

NOTES:

1 Definitions:

Maximum Contaminant Level (MCL) – The highest level of a containment that is allowed in drinking water. MCLs are set as close to MCLGs as feasible using the best available treatment technology and taking cost into consideration. MCLs are enforceable standards.

Maximum contaminant Level Goal (MCLG) – The level of a contaminant in drinking water below which there is no known or expected risk to health. MCLGs allow for a margin of safety and are non-enforceable public health goals.

Treatment Technique – A required process intended to reduce the level of a contaminant in drinking water.

2 Less than 5.0% samples total coliform-positive in a month.

3 MCLGs were not established before the 1986 Amendments to the Safe Drinking Water Act. Therefore, there is no MCLG for this contaminant.

4 Lead and copper are regulated by a Treatment Technique that requires systems to control the corrosiveness of their water. If more than 10% of tap water samples exceed the action level, water systems must take additional steps.

**Table E.2**
**National secondary drinking water requirements**

| Contaminant | Secondary Standard (mg/L) |
|---|---|
| Aluminum | 0.05 to 0.2 |
| Chloride | 250 |
| Color | 15 (color units) |
| Copper | 1.0 |
| Corrosivity | Noncorrosive |
| Fluoride | 2.0 |
| Foaming Agents | 0.5 |
| Iron | 0.3 |
| Manganese | 0.05 |
| Odor | 3 threshold odor number |
| pH | 6.5-8.5 |
| Silver | 0.10 |
| Sulfate | 250 |
| Total Dissolved Solids | 500 |
| Zinc | 5 |

# REFERENCES


Asano, T., Mujeriego, R., and Dickson Parker, J. November 1980. *Evaluation of Industrial Cooling Systems Using Reclaimed Municipal Wastewater, California State Water Resources Control Board*, Office of Water Recycling.

Cabelli, V.J. 1982. Microbial Indicator Systems for Assessing Water Quality. *Antonie van Leeuwenhoek,* 48:613.

California Department of Health Services. March 2001. *Recycling Water Quality.* Title 22 California Code of Regulations, Chapter 3, Section 60323.

Crook, J. 1991. Regulatory Issues Associated with Reuse Practices Throughout the World. *AWWA 1991 Annual Conference Proceedings, Resources, Engineering and Operations for the New Decade*. Philadelphia, Pennsylvania. June 23-27, 1991.

Crook, J., and Okun, D.A. Water Reuse: Past, Present, and Future. *AWWA Annual Conference Proceedings, Water Resources*. San Antonio, Texas. June 6-10, 1993.

D'Angelo, S., and Rice, K.C. 1984. Pilot Testing to Evaluate Virus Removal and Deactivation. *Proceedings of the 1984 Specialty Conference on Environmental Engineering,* ASCE/Los Angeles, California. June 25-27, 1984.

Deb, A.K., Weston, R.F., and Schorr, P. 1996. Industrial Water Reuse. *AWWA Annual Conference Proceedings, Water Resources Volume B*. Toronto, Ontario. June 23-27, 1996.

Florida Department of Environmental Protection (FDEP). August 1999. Reuse of Reclaimed Water and Land Application. *Chapter 62-610 Florida Administrative Code.*

Goldstein, D.J., Wei, I., and Hickes, R.E. 1979. Reuse of Municipal Wastewater as Make-up to Circulating Cooling Systems. *Proceedings of Water Reuse Symposium, Volume 1.* Awwa Research Foundation, Denver, Colorado.

Great Lakes Commission. 1996. *Liquid Asset: Great Lakes Water Quality and Industry Needs.*

McKee, J.E., and Wolf, H.W. 1963. *Water Quality Criteria.* Sacramento State Water Quality Control Board.

New Jersey Division of Water Quality. 2000. *Draft Technical Manual for Reclaimed Water for Beneficial Reuse. December 1999.* (Revised April 2000).

Office of Water Recycling. October–November 1978. *Industrial Water Reuse Conference Proceedings.* California State Water Resources Control Board, Office of Water Recycling.

Office of Water Recycling. March 1979. *Industrial Water Recycling.* California State Water Resources Control Board, Office of Water Recycling.

Office of Water Recycling. January 1982. *Water Recycling in the Pulp and Paper Industry in California.* California State Water Resources Control Board, Office of Water Recycling.

US Environmental Protection Agency. September 1988. *Constructed Wetland and Aquatic Plant Systems for Municipal Wastewater Treatment Design Manual*, United States Environmental Protection Agency (EPA), Office of Research and Development.

US Environmental Protection Agency. September 1992. *Guidelines for Water Reuse.* US Environmental Protection Agency Office of Technology Transfer and Regulatory Support. EPA/625/R-92/004.

US Environmental Protection Agency. 1998. *Water Recycling and Reuse: The Environmental Benefits*. EPA 909-F-98-001.

US Environmental Protection Agency. 2002. *List of Drinking Water Contaminants and Their Maximum Contaminant Levels.* USEPA Groundwater and Drinking Water. EPA 816-F-02-013.

US Geological Survey (USGS). 1995. *Estimated Use of Water in the United States in 1995.*

Utah Department of Environmental Quality Division of Water Quality. February 2001. Utilization and Isolation of Domestic Wastewater Treatment Works Effluent. *Title R317-1-4 Utah Administrative Code.*

Water Pollution Control Federation. 1989. *Water Reuse Manual of Practice.* 2nd ed.

Water Pollution Control Federation. 1990. *Natural Systems for Wastewater Treatment Manual of Practice FD-16.*

## ADDITIONAL READINGS

Barich, D.L. October 13-14, 1988. Health Criteria for Water Reuse. *Water Reclamation: Here, Now and How: Symposium III.* San Diego, CA.

Berry, J.L., and Kacmar, S.A. February 1-4, 1998. San Antonio Water System Water Recycling Program. *AWWA 1998 Water Reuse Conference Proceedings.* Lake Buena Vista, FL.

Crook, J., and Gorder, P. 1994. Water Quality Considerations for Non-Potable Water Reuse. *Proceedings 1994 Water Reuse Symposium.* Dallas, TX. February 27 - March 2, 1994.

Culp, G., Winser, G., Williams, R., and Hughes, Mark V. Jr. 1980. *Wastewater Reuse and Recycling Technology.* Park Ridge, New Jersey: Noyes Data Corporation.

DeBoer, J. 1984. Status of Water Reuse in the U.S.: An Overview of the 1984 Water Reuse Symposium III. *Proceedings AWWA Water Quality Technology Conference.* Denver, CO: AWWA. December 2-5, 1984.

Delaney, L.H. 1988. Water Quality Requirements for Water Reuse. *Water Reclamation: Here, Now...and How: Symposium III.* San Diego, CA. October 13-14, 1988.

Gregory, J.D. 1994. Revision of the National Reclaimed Water Guidelines. *Recycled Water Seminar Proceedings – Supplement Papers.* Newcastle, New South Wales, Australia. May 19-20, 1994.

Hartling, E.C. 1998. Overcoming Impediments to Water Recycling. *AWWA 1998 Water Reuse Conference Proceedings.* Lake Buena Vista, FL. February 1-4, 1998.

Kenhill Pty Ltd. 1999. *Industry Water Recycling Background Study.* Queensland Water Recycling Strategy.

Middlebrooks, E. Joe. 1982. *Water Reuse.* Ann Arbor, Michigan: Ann Arbor Science.

Spangenberg, C.W., and Ng, T. 1998. Conversion and Implementation of a Carpet Dye Manufacturing Process to Reclaimed Water: A Case Study. *AWWA 1998 Water Reuse Conference Proceedings*, Lake Buena Vista, FL. February 1-4, 1998.

Thakral, S.K., Lu, Y.W., and Kwan, S. 1996. Water Reuse in Pasadena, California - Its Challenges and Solutions. *1996 Water Reuse Conference Proceedings.* San Diego, CA. February 25-28, 1996.

# ABBREVIATIONS

| | |
|---|---|
| Al | aluminum |
| Ave | average |
| | |
| BDL | below detection limits |
| BOD | biological oxygen demand |
| | |
| $CaCO_3$ | calcium carbonate |
| CBOD | carbon biological oxygen demand |
| Cl | chlorine |
| COD | chemical oxygen demand |
| Con | consecution |
| | |
| DO | dissolved oxygen |
| | |
| Fe | iron |
| | |
| gm | geometric mean |
| | |
| Max | maximum |
| Med | median |
| mg/L | milligrams per liter |
| Min | minutes |
| ml | milliliter |
| MPN | most probable number |
| | |
| N/A | not applicable |
| $NH_4$-N | ammonia, nitrogen |
| NTU | nephalometric turbidity units |
| | |
| Org | organism |
| | |
| PAC | Project Advisory Committee |
| $PO_4$ | phosphate (phosphorus oxide) |
| | |
| $SiO_2$ | silicate (silicon oxide) |
| | |
| TDS | total dissolved solids |
| Temp | temperature |
| TOC | total organic carbon |
| TOX | total organic halogen |
| Tr | trace |
| TSS | total suspended solids |
| | |
| USEPA | United States Environmental Protection Agency |

IWA Publishing's authorised EU representative for General Product Safety Regulations is Diane D'Arras, 15 rue Duret, 75116 Paris, France, e-mail: safety@iwap.co.uk.

Printed and bound by CPI Group (UK) Ltd, Croydon, CR0 4YY
17/09/2026
02218813-0014